who should play GOD?

Other Books by
Ted Howard and Jeremy Rifkin

Common Sense II
Own Your Own Job

who should play GOD?

The Artificial Creation of Life and What It Means for the Future of the Human Race

Ted Howard and Jeremy Rifkin
of the Peoples Business Commission

DELACORTE PRESS/NEW YORK

To Aldous Huxley
he foresaw

Published by
Delacorte Press
1 Dag Hammarskjold Plaza
New York, New York 10017

ISBN: 0-440-09552-2

Manufactured in the United States of America
First printing

ACKNOWLEDGMENTS

This book was a project in two stages: the first, research; the second, writing. Without the first, there could not have been the second. Dan Smith and Larry Gordon are largely responsible for gathering together the massive amount of historical and scientific information that forms the basis of this book. Their many months of long days spent in libraries, at professional associations, sitting in on meetings, and conducting interviews were critical in shaping our own arguments and conclusions regarding the Biological Revolution. We gratefully acknowledge their contribution to this book.

We would also like to thank Mary Murphy, Randy Barber, and Noreen Banks for their assistance and support. Thanks, too, to Martha Kinney, our editor, for her understanding of, and commitment to, this project.

Finally, we want to acknowledge Science for the People, a nationwide organization of scientists who have devoted their lives to insuring that science and technology serve humanity rather than enslave it. Their contribution toward building a humane and democratic America should be remembered when the history of our era is recorded.

CONTENTS

INTRODUCTION

Life has existed on earth for three billion years. In that time over ninety-eight million different species have come and gone.

The newest organism to appear in this evolutionary chain is Homo sapiens. In the total history of life on the planet earth, human existence accounts for less than three million years. In contrast to bacteria, which have gone through ten trillion generations, shelled sea fish, which have gone through six hundred million generations, and horses, which have gone through twenty-seven million generations, humans have lived through only two hundred thousand generations.[1]

In fact, if you were interested in tracing your direct family tree back to the height of early Greek civilization, you would find a mere one hundred generations separating you from the days of Aristotle, Plato, and Socrates.

In this rather short period of our history, there have occurred singular moments in time when the body of accumulated knowledge spawned dramatic new inventions and discoveries which humans in turn have used to regulate, control, and moderate the external world around them: for example, the discovery of fire, the invention of the wheel, the formulation of written language, the discovery of the principles of gravity, and the development of the internal combustion engine. Human history up to now has been the history of constantly tailoring and retailoring the outside environment to suit specific needs. The splitting of the atom and the dawn of the nuclear age are the latest developments in this long attempt to harness nature.

Throughout this unfolding process, which we call civilization, scientific and technological progress has never been equally distributed. New discoveries have always been applied selectively, with some group, class, or race using knowledge of the external world to control not only it but their fellow humans as well. As the late C. S. Lewis observed, "Man's power over nature is really the power of some men over others with nature as their instrument."

Now a dramatic new scientific discovery has given some people the power, for the first time, to shift attention from shaping and controlling the external world of matter and energy to shaping and controlling the internal world of life itself. With the discovery of DNA and its workings, scientists have unlocked the very secrets of life. It is now only a matter of a handful of years before biologists will be able to irreversibly change the evolutionary wisdom of billions of years with the creation of new plants, new animals, and new forms of human and post-human beings.

Today, only a tiny handful of people are privy to the secret of life and how to manipulate and change it. Most people are totally unaware of this new-found power. The concept of designing and engineering life, especially human life, is so utterly fantastic that it is difficult even to comprehend its meaning and implications. Yet, even as the public is kept virtually ignorant of this unparalleled new scientific discovery, microbiologists are busy at work in hundreds of laboratories across the country, spending tens of millions of dollars in pursuit of the "mastery of life."

This book takes a look into these laboratories and the ideas, motivations, and objectives of those who are sponsoring and conducting research into the redesigning of life.

What you are about to read is not science fiction. Genetic engineering (the artificial manipulation of life) is a very real phenomenon—so real, in fact, that it has sparked an unparalleled controversy within the scientific community itself. Chapter I provides a brief overview of the field of genetic engineering and the controversy surrounding it. Chapter II goes on to explore the social philosophy underlying genetic research. The history of eugenics, the ideological wing of genetics, has been virtually written out of popular American textbooks. Chapter II traces that history from its earliest beginnings to today and includes the arguments currently being set forth by modern eugenicists to secure wide acceptance for mass human genetic engineering. Chapters III and IV are a look into the laboratories and an examination of the new discoveries that are making possible everything from test-tube babies to cloned humanoids and chimeras. Chapters V and VI describe the ways in which human genetic engineering is likely to be introduced into society, and the people and institutions that will be promoting its applications. The concluding chapter raises some of the basic issues and questions posed by genetic engineering and makes some concrete suggestions on what should be done about this research.

For many years social commentators have looked on nuclear weaponry as the most powerful and dangerous tool at the disposal of humanity. With the development of human genetic engineering, a tool even more awesome is now available. It is true that nuclear weaponry poses the ever-present threat of annihilation of human life on this planet. But with genetic engineering there is a threat of a very different kind: that by calculation and planning, not accident or the precipitous passion of the moment, some people will make conscious and deliberate decisions to irreversibly alter the biological structure of millions of other men and women and their descendants for all time. This is a form of annihilation every bit as deadly as nuclear holocaust, and even more profound—whatever forms of future beings are developed will be forced to live the consequence of the biological designs that were molded for them.

It should be said at the outset that this book is not intended to be value-free in examining what we regard to be the most important single question ever to face humanity. The question is whether we should preserve the human species and other forms of life as they are presently constituted or forge ahead on a mass program of biological reengineering over the next twenty-five to fifty years. On this question we side with the opponents of genetic engineering, and this book is intended to reflect that point of view.

1
the biological revolution

IN THE BEGINNING . . .

Recombinant DNA. A few short years ago, those words had meaning only for a handful of molecular biologists. Today they signify what is fast becoming a heated social controversy that seems destined to eclipse public concern over the atom bomb.

A Nobel Prize–winning biologist warns that this astounding new technology "faces our society with problems unprecedented not only in the history of science, but of life on the Earth."[1] The chairman of the microbiology department at a leading university claims that it is "the most powerful tool to come along in my lifetime."[2] One of the nation's most prominent religious figures fears that when applied to human beings, recombinant DNA may lead to social abuses comparable to "another Dachau."[3] Another scientist warns that with this discovery, "only one accident is needed to endanger the future of mankind."[4]

Discovered only in 1973, recombinant DNA—a laboratory technique for splicing together genetic material from unrelated organisms to manufacture novel forms of life—raises the most significant ethical, political, and social dilemmas a society has ever had to face. While the scientific community has become deeply divided over the health and safety problems presented by recombinant DNA, lay observers are increasingly questioning how this new technology will be used to modify human life as we know it. As a tool of human genetic engineering, recombinant DNA literally offers us the opportunity to move beyond ourselves on the evolutionary scale. "Man the engineer" may soon become "man the engineered."

The implications of recombinant DNA are so staggering and unprecedented that the technique itself seems to have burst upon the world like a comet from the far reaches of space. But in fact, the development of recombinant DNA fulfills a twenty-five-year-long search by thousands of biologists intent upon the mastery of the human genetic process. Beginning in the early 1950s, an explosion of basic knowledge has dramatically altered our understanding of life. Today, molecular biologists probe and analyze the basic chemical substances of living matter. Experimentation now proceeds in the microscopic world of the gene, the unit that determines heredity. Geneticists gaze into the very mysteries of life, searching for the keys that will one day unlock the doors to the biological control of the future of humanity. Many have already been found. To Harvard Nobel laureate Salvador Luria, "the relevant point . . . is that all the essential features of the genetic process, insofar as they have been clarified, have turned out to be interpretable in strictly biochemical terms. What molecular biologists have done is to make the genetic mechanism directly available to chemical experimentation."[5]

Like the atomic bomb, the powerful new "tools" for understanding the chemical components of life are sure to wield a tremendous influence on our future. "It is clear," says Dr. James F. Crow, chairman of the biology department of the University of Wisconsin, "that biological and chemical possibilities for influencing human evolution and development are certain to come, probably before we have thought them through."[6] A report issued by the Subcommittee on Science Research and Development of the House of Representatives is even more explicit: "The science of genetics is rapidly moving out of the realm of theoretical research and into the more politically sensitive region of applied science. The technological capability to alter the course of human evolution is relatively close at hand."[7]

Where that new course will ultimately lead is the subject of much discussion and conjecture among the bioengineers. The goals of today's genetic engineers range from the sublime to the ridiculous. Still, no matter how wild or fantastic the genetic scenarios they envision may be, it is clear that those who are doing the proposing are deadly serious. The most modest among them advocate applying genetic engineering to curing some two thousand monogenic diseases—disorders caused by the malfunctioning of a single gene. Others are turning their attention to the artificial production of new strains of plants and "super" grains, while still others focus on redesigning various animal species.

More ominous are the well-credentialed and well-financed researchers who propose the complete retailoring of human life. Name your wildest fantasy, or nightmare, and some authority somewhere is seriously proposing it: from redesigning human stomachs so that people will be able to consume cheap hay and grass, like cows, to the hybridization of humans with lower primates.

There are even some genetic engineers who eagerly await the day when their work will produce the "final solution": the construction of a genetic super-race that will move far beyond Homo sapiens on the evolutionary ladder.

Once, all of this could be dismissed as science fiction, or the mad ravings of a Dr. Frankenstein. No more. We are not in the Brave New World yet, but we are well along the road. Today we live in a new era, a new reality. Today we live in the midst of the Biological Revolution.

THE DISCOVERY OF DNA

Like all of the great radical movements of history, the Biological Revolution embodies its own mythology, complete with founders, important dates and events, even its own radical tract that proclaimed to the world that the break with the past had begun.

The year was 1953, the place a small "hut" in Cambridge, England. Two highly ambitious and imaginative scientists, James Dewey Watson, a twenty-five-year-old American, and Francis Crick, a British physicist-turned-biologist, huddled over a tinkertoy-like model of spidery arms and interlocking metallic clamps. Their excitement was obvious, for they were in the presence of something that no one had ever before seen. By lunchtime, Watson later recalled, Crick was seated in a dining hall telling "everyone within hearing distance that we had found the secret of life."[8] A few weeks later, Watson and Crick made the formal public announcement of their discovery. It came in the form of a barely one-thousand-word article published in the April 25 edition of the British science magazine *Nature*. For the brash young duo, the document's opening sentences are uncharacteristically modest: "We wish to suggest a structure for the salt of deoxyribonucleic acid (DNA). This structure has novel features which are of considerable biological interest."[9]

That announcement transformed the field of biology. Many immediately hailed it as the biologists' equivalent of the physicists' splitting of the atom. Dr. Peter B. Medawar, the British Nobelist, has called it "the greatest achievement of science in the twentieth century."[10] It was certainly one of the epic discoveries in the history of scientific thought.

What Watson and Crick found, through a process of experimentation, molecular model building, X-ray photography, good guesswork, and the accumulated research of many other scientists, was the physical make-up of DNA, the fundamental molecule of life. It was, as Watson later wrote, "a beautiful new structure"[11] because of its very simplicity. Located along the chromosomes of the cell, DNA takes the shape of a long, twisting *double helix,* a kind of microscopic spiral staircase. The steps of the staircase, called *base units,* are the most significant component of the structure. Composed of just four different chemical nucleotides—adenine, thymine, guanine, and cytosine—the base units arrange themselves in an endless variation of patterns that form the genes. The collective interaction of the genes determines the production of various proteins, which in turn direct the growth and development of living organisms. String a few hundred genes together and the result might be a very simple life form—bacteria, for example. Hook a million genes together, and something else will result—a man or woman, for instance. Or an elephant, an oak tree, or a polar bear.

Conceptually, DNA is a tiny computer, miniaturized beyond Sony's wildest dreams. If all the DNA in the cells of a human being were unraveled and strung together, it would stretch the distance from the earth to the sun and back four hundred times. Yet, in its natural state, the DNA contained in a human being is so tightly packed that it could fit into a cube measuring just one square inch on a side.[12]

Programmed into this microscopic data bank is an almost inconceivable amount of information that spells out the biological and chemical functions of life. A single thread of DNA located in just one human cell may house as much information as one thousand books, each six hundred pages thick. A teaspoon of DNA taken from simple bacteria has been equated to the storage capacity of a modern computer, one hundred cubic miles in volume![13]

All of these bits of DNA information spell out the critical instructions that determine everything from the color of your eyes and the numbers of toes and fingers to the size of the brain. It is these bits of genetic data that molecular biologists are now learning to read. Once having learned the letters of the alphabet, they will ultimately be able to write.

THE LAST THREE BILLION YEARS

DNA is the most universal of languages; it is shared by all life forms that exist today, and those that ever existed in the past. Dr. Charles Price, former president of the American Chemical Association, hypothesizes that the fact that the "DNA language in the simplest one-cell organism is exactly the same DNA language that duplicates a human being . . . suggest[s] that one original cell became the progenitor of all life on earth."[14]

For the genetic engineer, that is a heady thought. It means, quite simply, that the Biological Revolution, with its ability to manufacture and alter DNA information, stands in command of the evolutionary past and future. No modification of any life form is outside the eventual technological capability of the new genetics. We have already progressed to the point, says Dr. Robert Sinsheimer, chairman of the biology department of the California Institute of Technology, where genetic engineering "makes available to us the gene pool of the planet—all of the genes developed in the varied evolutionary lines throughout the history of life—to reorder and reassemble as we see fit."[15]

The natural evolutionary process is the result of

painstakingly slow and careful mutations of the genetic material of organisms. In the past, it has taken anywhere from four million to twenty million years for a single mutation to establish itself throughout the entire population of a species. Now, with the Biological Revolution upon us, scientists are proposing massive genetic intervention virtually overnight. Dr. James F. Danielli, an advocate of what he calls "The Age of Synthesis," happily predicts that he and his colleagues will soon speed up nature's way of doing things to the magnitude of one billion times a year.[16]

One need only look at the history of life on this planet to grasp the enormity of the genetic alterations that are being proposed. Jacques Yves Cousteau has suggested that "if distance is compared with time, a reasonable appreciation of the biological process can be obtained."[17] Suppose, for example, that one step is equivalent to fifty years. Four steps would take us back to the American Revolution. Forty steps take us back to the time of Christ, two hundred to the time when prehistoric people sought shelter in caves. Fifteen miles back, our ancestors appeared. But it would involve a journey of some twenty-five thousand miles—the distance around the earth—to arrive at the starting point of the evolutionary process.

It all began over three billion years ago, in the primordial soup of the ocean. Something happened—cosmic radiation, perhaps, or tremendous electrical sparks—that caused inorganic gases and chemicals to come together and form life. At first the organisms were barely on the edge of the living: bacteria and algae. But since that time, the rules of the game have remained the same: life growing and evolving, becoming increasingly complex, rising higher on the evolutionary scale through rare, but critical, mutations in genes.

The awesome slowness of the process is difficult to comprehend. For 2.5 billion years, these simple life forms were all that existed amid the life-sustaining water. Then, 600 million years ago, the first invertebrates—forerunners of jellyfish, sponges, worms, and snails—began to take shape from the bacteria and algae.

Another 150 million years passed before the next major evolutionary leap, the development of fish and other vertebrates. Fifty million years later, another development burst forth: Fish began to grow jawbones, making them the first creatures capable of biting and eating other life forms. More important (from our perspective on the evolutionary ladder, at least), a new type of life, the lungfish, emerged. Capable of breathing air outside of the water, and blessed with muscled fins that could be used for walking, these lungfish eventually left the ocean to explore the land.

Next, the amphibians, 350 million years ago. Along the shores of steamy swamps the ancestors of present-day frogs, toads, and salamanders huddled among newly developed plants, first in the form of moss, later as exotic ferns and flowering flora.

The evolutionary process continued. Reptiles: snakes, lizards, crocodiles, turtles, their skin mutating into scaly hardness to permit survival in drier climates. Even their eggs changed over millions of years, developing hard shells to permit nesting on the land.

Then came the dinosaurs, seventy-foot-long, eighty-five-ton behemoths that roamed the land, dominating the world for 150 million years. Next, the age of the mammals, a new life form adapted to a drying and cooling climate, a climate that spelled extinction for the dinosaur. Primitive cats and dogs, elephants, horses, and pigs appeared. Grain plants flourished; fruits and grasses developed.

Finally, just 2.5 million years ago, our closest ancestors, "prehistoric man," emerged from the apes. Not until 35,000 B.C. did "modern man" appear. More than thirty thousand years later, what we call "civilization" finally began to take hold.

Over these past three billion years, one hundred million species have existed on this planet. Of those, ninety-eight million are now extinct. Among the two million that remain today, only one, Homo sapiens ("wise man"), has evolved to the point of being able to control and harness its own evolutionary future. Many biologists welcome this possibility, seeing it as a great challenge that will ennoble and preserve our species. "Modern progress in microbiology and genetics suggests that man can outwit extinction through genetic engineering," argues Cal Tech biologist James Bonner. "Genetic change is not basically immoral. It takes place all the time, naturally. What man can do, however, is make sure that these changes are no longer random in the gigantic genetic lottery in nature . . . Instead, he can control the changes to produce better individuals."[18] Bonner's viewpoint is seconded by Dr. Joseph Fletcher, professor of Medical Ethics at the University of Virginia School of Medicine, who sees in genetic engineering the fulfillment of our cosmic role on earth. "To be men," he believes, "we must be in control. That is the first and last ethical word."[19] Promises a third scientist, our newly developed eugenic potential will lead humanity to "a growth of social wisdom and glorious survival—toward the evolution of a kind of superman."[20]

But what, the nonbeliever might ask, constitutes a "superman"? Who will decide on the traits and attitudes these "better individuals" embody? Who among us will be in control of securing our evolutionary future, and the evolutionary future of the other animals and plants that will be subject to genetic engineering? And how can we be sure that, in our tinkering and fixing with three billion years of evolutionary wisdom, we do not inadvertently join the ninety-eight million species that have passed from the earth?

THE BIOLOGICAL REVOLUTION

The last quarter century is but a speck of time on the three-billion-year continuum of evolution. The past twenty-five years, however, have produced an explosion of information that lays bare many of the secrets of those hundreds of millions of years. This explosion is sure to continue. Biological knowledge is currently doubling every five years; in the field of genetics, the quantity of information doubles every twenty-four months.[21] One geneticist estimates that the serious molecular biologist must now read a minimum of one hundred thousand pages of articles in scholarly journals just to keep up on the most recent developments.[22]

Dr. Sydney Brenner, of the Medical Research Council Laboratory of Molecular Biology in Cambridge, England, traces this explosion directly to the atmosphere of excitement that surrounded the work of Watson and Crick. "The double helix fundamentally changed the image of biology," he believes. "To most young people of my generation, the biology taught in universities was a most unattractive subject. It seemed to consist of learning long dusty lists of Latin names punctuated by cutting up frogs or carrots in long dusty laboratories. DNA changed all of that and turned biology into an exciting, intellectually attractive subject."[23]

Many of those attracted to the exciting new biological frontier, like Francis Crick, had formerly conducted research in the field of physics. Morally repelled by the development of nuclear weapons as an outgrowth of their work, physicists turned to the "life-giving" science of biology. Ironically, they used their knowledge of physics to transform old-fashioned biology into molecular biology, a field that would soon come to embody the very same ethical and political considerations as the use of the atomic bomb. As the Biological Revolution gains momentum, many scientists have come to share the sentiments of one prominent biologist who notes, "I feel closely akin to the physicists who pointed out in the 1930's that the principles required for the release of the energy locked in the atomic nucleus were understood. All that we lacked was a practical breakthrough and the requisite technology. Here, too, the principles seem in hand. All that seems really needed is optimism, sustained effort, and support commensurate with the importance of the problem."[24]

That support, in the form of a massive expansion of government financing of science, provided the material base for the rapid growth of molecular biology, says Brenner. "Watson and Crick may have invented it, but Uncle Sam certainly fuelled it."[25]

The infusion of new talent and big money, including the increasing backing of corporate America, has produced startling results. Summing up the field of biology in recent years, a 1974 government report concluded: "The advances which have taken place . . . , while poorly understood by the general public, have been nothing short of miraculous in terms of scientific progress, exceeding, in the opinion of many investigators, the frequently cited parallel developments in nuclear physics."[26]

Even a brief overview of the progress made in the field of molecular biology during the past quarter century is enough to boggle the mind. Each discovery, each revelation, each new piece of the genetic puzzle has had an effect magnified many times beyond itself. The accumulation of genetic knowledge continues steadily and unrelentingly. With many thousands of researchers at work, the borders of the unknown are pushed inexorably back.

Consider, for example, just ten of the scores of major advances made over the last two decades:

• *We have learned how DNA reproduces itself.* In 1957, Dr. Arthur Kornberg, performing a series of test-tube experiments, determined that the structure of DNA replicates by "unzipping" itself. Once separated into two strands, the DNA then attracts new chemical substances from the surrounding cell, making an exact duplicate of its original structure. Since Kornberg's original revelation, we have learned how to reproduce copies of DNA artificially in a form clearer and more stable than the original.[27]

• *We have cracked the DNA code.* The base units along the double helix form into what are called *DNA triplets,* a series of three-letter words. There are sixty-four possible combinations of these words, each triplet coding the instructions for production of a specific protein. Large quantities of proteins, in combinations, produce the variations in organisms. In 1961, Dr. Marshall W. Nirenberg performed the biological equivalent of deciphering the Rosetta stone by isolating a DNA triplet and determining the protein it produced. Quickly, other biologists have managed to analyze the entire sixty-four-word DNA code.[28]

• *We have learned how DNA transmits its instructions to the cell.* Even with Watson and Crick's discovery of the structure of DNA, scientists did not understand how a gene in a DNA molecule sends out its chemical instructions for the building of proteins. Then, about 1960, several teams of researchers in the United States and France isolated *messenger RNA,* the mechanism which acts as the DNA information carrier.[29] Since that first discovery, other forms of RNA have been found, and their purpose determined. In April of 1977, a team of scientists at the University of California at San Francisco announced that they had successfully "ordered" messenger RNA to reproduce a strand of the original DNA that had dispatched it.[30]

• *We have analyzed chromosomes to determine genetic function.* In 1956, researchers determined that the human cell carries forty-six chromosomes, packages of DNA on which the genes are located. A chromosome sample can now be taken from a three-month-old fetus, and through photographic analysis the possible occurrence of some sixty genetic diseases can be predicted. Chromosomes are also being subjected to laser-beam surgery, in which one part of a chromosome is sliced from the whole. By determining how the entire cell reacts when specific pieces of individual chromosomes are destroyed in this way, scientists expect to learn all of the different functions of each chromosome. Still to be used in chromosomal research is a special electron microscope that can measure an object four billionths of an inch in size. Dr. Albert V. Crewe, one of the developers of this technique, believes that one day it will be used to literally read off the genetic code along a strand of the chromosome.[31]

• *We have synthesized a cell.* Dr. James F. Danielli headed the team that first built a cell. They accomplished the feat by reassembling parts of three different amebas into one functional whole.[32] Scientists have also produced "giant cells" with the ability to grow five hundred to a thousand times larger than normal. Likewise, mini-cells have been engineered.[33]

• *We have fused cells from two different species.* We can now take cells from two different organisms—a mouse and a human, say—and fuse them together, producing a hybrid cell that carries some of the properties of the two originals. Cell fusion has also been accomplished between human and plant cells, hens' red blood cells and yeast, and carrot cells and cells taken from a human cancer victim.[34]

• *We have isolated pure human genes.* In 1969, Dr. Jonathan Beckwith and a team of researchers at Harvard reported the first isolation of a pure gene from bacteria.[35] By 1973, the purifying of the first human gene had been accomplished. With the gene isolated from varied chemical reactions that take place in the cell, biologists can analyze it in a test tube. One outgrowth of this work is the increased understanding of the mechanism that turns genes off and on.[36]

• *We have "mapped" genes.* The genes responsible for various physical traits—hair color, for instance—are located at specific locations on specific chromosomes. Scientists are learning to locate, or map, genes on the chromosome. The mapping process began in the early 1970s; today, well over two hundred genes have been mapped.[37]

• *We have synthesized a gene.* During the last decade, Dr. Har Gobind Khorana of MIT has pioneered in the synthesis of genetic material. In 1970, Khorana succeeded in synthesizing a gene found in a yeast cell. By 1976, he had performed a considerably more dazzling trick in building from scratch, using only the basic four nucleotides, a two-hundred-sequence human gene. As an added feature, Khorana equipped his creation with the start and stop mechanisms critical to controlling its function. Once inserted into a cell, the synthetic gene proceeded to work perfectly.[38]

• *We have changed the heredity of a cell.* In 1971, a team of scientists introduced a gene from a bacterium into a human cell where it functioned, thus changing the original instructions received by the cell.[39] In another highly significant experiment involving cell heredity, cells from one mouse were injected into the fetuses of another. Not only did the genetic material from the first mouse show up in the newborn mice, but when these mice mated and had offspring, the foreign genes were passed on to the next generation. Genetic alteration had been permanently incorporated into the germ line of a mammal, a development that one scientist believes "brings us ten years closer to the possibility of genetic engineering of humans."[40]

Behind these technological triumphs lurks an awesome truth. In essence, the chemistry of the gene now stands revealed. As with so many other technologies, each new bit of genetic knowledge opens the way to major innovation. In itself, for instance, the ability to fuse cells of two different species is fascinating. The importance of the experiment, though, lies not in the marvel of the technique, but in the end product. When a human cell and a mouse are fused together, some of the chromosomes drop away. Each time the fused cell divides, additional chromosomes disappear. The researcher can use this technique to determine the function of the genes on the chromosomes that do not duplicate. Cell fusion, says Dr. Hayden Coon of the National Institutes of Health, "makes it possible to map every testable function of human cells and to find out on which chromosome each controlling gene is located."[41] When a line of research like cell fusion is coupled with other genetic technology—the synthesis of genes, say, or the ability to introduce functioning genetic material into cells—the possibility of applied eugenics (or, as some euphemistically call it today, "genetic intervention") becomes a technical reality.

This is the real significance of the myriad of genetic discoveries of the Biological Revolution. In the context of the entire field of molecular biology, even the most esoteric experiments are imbued with vast importance. "Looking back," declares Sydney Brenner, "one can see that the double helix brought the realization that information in biological systems could be studied much in the same way as energy and matter." Genes, he says, were turned into "chemical objects whose structure and function could be analyzed and understood in terms of biochemical machinery."[42]

The metaphor of the gene as machine is one that is used consistently by genetic engineers. The prevailing attitude was summed up by one scientist who announced to his fellow researchers, "our job is simply to find out how these interesting pieces of machinery work, how they get built and how they came to be the way they are . . . That is why molecular biology seems to me to be the art of the inevitable."[43] From the outset of the Biological Revolution, the very conceptual simplicity of the structure of DNA has proved seductive to those who would shape and redirect life. Now that seductive possibility is at hand. By the end of the century, suggests Nobelist Marshall Nirenberg, the biologist who cracked the DNA code, we will be able to program into the code "synthetic" messages, turning on some genes, turning off others, rearranging genetic patterns, manufacturing life.[44]

The cumulative effect of the recent events in molecular biology has caused noted British author and science observer Lord Ritchie-Calder to worry that "we are entering an Age of Bioengineering . . . Just as we have manipulated plastics and metals, we are now manufacturing living materials and the very nature of human beings."[45] In a technological society such as ours, this kind of manufacturing process is not an aberration; it is a logical extension of scientific discovery. Whenever a science develops to the point of generating new technology, says Dr. Salvador Luria, the question that faces society is no longer feasibility. Once the scientific principles are established, "technological application is almost certain to come."[46]

That certainty has now become real, not only in the minds of the scientists, but for laymen as well, with the discovery of the incredible new technique of recombinant DNA, a process which, *Science* magazine says, "is widely expected both to revolutionize the study of molecular biology and to bring some of the wilder fantasies of genetic engineers into the realm of the possible."[47]

RECOMBINANT DNA— MODERN-DAY CHIMERAS

Among the creations of Greek mythology was the Chimera, an animal with the head of a lion, the body of a goat, and the tail of a dragon. The monster vomited forth horrible flames, and, the myth has it, personified a terrible storm—"pouncing suddenly on the darkened waves, unleashed the raging tempests to destroy men."[48]

Twenty-five hundred years after the flowering of Greek civilization, two teams of scientists led by Drs. Stanley Cohen and Herbert Boyer reported that they had constructed a novel life form. Taking two unrelated organisms that would not mate in nature, Cohen and Boyer isolated a piece of DNA from each and then hooked the two pieces of genetic material together. The result was literally a new form of life, one that had never before existed on the face of the earth. They dubbed their creation "DNA chimeras," because, Cohen says, "they were conceptually similar to the mythological chimera."[49] Cohen, of course, was referring to their hybrid nature. He did not know at the time that his chimera—or, as it is technically known, recombinant DNA—was destined to unleash "raging tempests" of its own, provoking one of the bitterest battles in the history of modern science, and perhaps the most important.

The discovery of the recombinant DNA process in 1973 was immediately hailed by biologists with a zeal equaled only by their earlier praise of Watson and Crick's discovery of the structure of DNA. With the new techniques, the genetic material from virtually any two unrelated organisms can be spliced together, opening up a whole new arena of experimentation and study. "Now and in the next few years we shall be doing things that would have been thought completely improbable a few years ago," says Dr. Paul Berg of Stanford.[50] Another researcher predicts that "problems people were interested in 100 years ago are starting to be tractable now with these methods."[51] A third, Dr. Oliver Smithies of the University of Wisconsin, sees the discovery of recombinant DNA as "one of the most significant advances of 20th century biology."[52]

A product of nearly thirty years of investigation, climaxed by a series of rapid discoveries in the late 1960s and early 1970s, recombinant DNA is a kind of biological sewing machine that can be used to stitch together the genetic fabric of unrelated organisms. Dr. Cohen sees it as genetic surgery in four stages. To begin with, a chemical scalpel, called a *restriction enzyme,* is used to split apart the DNA molecules from one source—a human, for example. Once the DNA has been cut into pieces, a small segment of genetic material—a gene, perhaps, or a few genes in length—is separated out. Next, the restriction enzyme is used to slice out a segment from the body of a *plasmid,* a short length of DNA found in bacteria. Both the piece of human DNA and the body of the plasmid develop "sticky ends" as a result of the slicing process. The ends of both segments of DNA are then hooked together, forming a genetic whole composed of material from the two original sources. Finally, the modified plasmid is used as a *vector,* or vehicle, to move the DNA into a *host cell,* usually a bacterium. Absorbing the plasmid, the bacterium proceeds to duplicate it endlessly, producing identical copies of the new chimera. These are called *clones.*

On the surface, the combining of different organisms in this way—mouse DNA with human, or rhododendron with horse—may seem bizarre. Researchers, however, are quick to point out that they are not creating giant mice with human heads that will walk out of a laboratory à la Frankenstein's monster. The purpose of recombinant DNA is considerably more significant than that. What the new technique makes feasible is the isolation and purification of a gene in massive quantities. With dozens of different types of restriction enzymes now available, researchers can slice DNA molecules in specific spots to isolate out specific genes. In May of 1977, for instance, a research team in California announced that, using the technique, it had isolated and analyzed a gene that orders the production of the hormone determining human growth. Now that this gene has been scrutinized, the next step will be to implant it in a human cell and stimulate it into functioning. Researchers say this technique will ultimately provide a cure for dwarfism. Just as easily, it could be used to develop a basketball team of ten-footers.[53]

Though recombinant DNA is still in its tooling-up phase, its proponents already envision a wide range of applications that will solve various medical and social problems. Cohen, for one, imagines the synthesis of "a variety of biologically produced substances such as antibiotics and hormones, or enzymes that can convert sunlight directly into food substances or usable energy."[54] Lederberg anticipates "the production of high-quality food protein supplements."[55] Others have volunteered scenarios in which recombinant DNA solves the nation's fuel crisis by converting waste into energy, retrieves precious minerals from the ocean, cleans up oil and chemical spills, and eliminates worldwide pollution.

In the medical realm, Lederberg believes, recombinant DNA may one day become the ultimate drug that will provide humanity with "some of the most pervasive benefits for the public health since the discovery and promulgation of antibiotics."[56] Dr. David Baltimore, a Harvard Nobel laureate, carries the claims a step further. "How much do we need recombinant DNA?" he asks rhetorically. "Fine, we can do without it. We have lived with famine, virus and cancer, and we can continue to."[57]

There are also the farther-reaching applications of recombinant DNA. Those biologists who have been the strongest proponents of human genetic engineering are also those who have pioneered in the development of this new technology. Recombinant DNA is "an extraordinarily powerful and simple tool for studying the structure and function of mammalian DNA," says Harvard's Bernard Davis, "and it has rapidly become as indispensable as radioactive isotopes or the electron microscope."[58] For the researcher intent on analyzing human gene function, recombinant DNA is indeed becoming the greatest shortcut available in molecular biology.

The work of two scientists, W. Gilbert and F. Sanger, provides dramatic testimony to just how significant that shortcut is. In the past, Gilbert and Sanger spent as long as two years determining the sequence of genes along a two-hundred-base-unit fragment of DNA. Using the recombinant technique, they can now determine such a sequence in just two days.[59] Eventually, recombinant DNA will be used to determine the sequence and function of all one hundred thousand human genes. Once that has been accomplished, the final hurdle will be making recombinant DNA not only copy itself when placed inside a cell, but take command of the functions of the cell as well. When this final barrier is breached, we will truly be on the road to the total control of our evolutionary future.

CHIMERA OR TROJAN HORSE?

So great are the expectations surrounding recombinant DNA that a major scientific/corporate/government complex has sprung up around it, a technological boom town set to mine the new genetic mother lode. The federal government has become an enthusiastic supporter of the research, handing out at least 180 recombinant DNA research grants to over 80 laboratories around the country in one year alone.[60]

Seven major pharmaceutical companies are also engaged in the recombinant race, and a dozen more drug, chemical, and agricultural companies are poised to enter a field which *Fortune* magazine predicts will soon become a "multi-billion-dollar industry."[61] Researchers at General Electric have already developed a recombinant oil-eating microorganism that they hope to use one day to clean up oil slicks. (Whether this oil guzzler will be content to munch on oil spills, or will go on to pillage automobile gas tanks and eventually the Texas oil fields, no one can predict.) The U.S. Patent Office has granted GE three patents for its little bug, and at one point announced that it would even speed up the processing of patents involving the recombinant technique so that those who play God with new life forms can have exclusive ownership to them.

DNA has clearly come of age. Summing up progress in recombinants, a Swiss scientist observed during an international genetics conference, "this sort of research is gaining uncontrollable momentum."[62]

But now the catch. Like another Greek mythical figure, the Trojan horse, recombinant DNA chimeras hide within their outward beauty a deep and dark danger, one that may be as destructive to humanity as the horrors of nuclear holocaust. In splicing together genes from two different species, scientists are creating novel forms of life whose properties they cannot know beforehand. Once the two kinds of DNA are joined together, the chimera is placed within a host, usually the bacteria strain *Escherichia coli,* where it quickly proceeds to copy itself. *E. coli* has been selected by most researchers to house the recombinant DNA simply because more is known about it than about any other bacteria on the planet. There is a good reason for our encyclopedic knowledge of *E. coli*—it is a common inhabitant of human intestines. And that is where the problem comes in.

What might happen, researchers have begun to wonder, if a recombinant DNA chimera resulted in a cancer-causing alignment of genes? Placed in the protective environment of the *E. coli,* the new, cancerous life form would churn out cancer like a Detroit factory stamping out autos. If this *E. coli* found its way out of the laboratory, carried on a lab technician's skin, it would then roam free among trillions of its friends, and likely would pass some of its cancer around. Once these virulent *E. coli* began invading human systems . . . well, no one knows exactly what might happen. A new plague? A cancer epidemic? A real-life Andromeda Strain?

Even a seemingly benign recombinant DNA creation might ultimately have serious consequences. Dr. Paul Berg of Stanford, himself a leader in the recombinant DNA field, admits that "it is possible that such infections might not produce immediate symptoms, and therefore it might be years before harmful effects of a tumor virus became evident."[63] The result would be a biological time bomb, waiting to go off.

To add to the problem, restriction enzymes necessary for cutting apart DNA molecules are available commercially, sent through the mail to anyone who requests them. Science magazines abound with ads for these enzymes and the other chemicals and equipment necessary to set up a recombinant DNA shop right in your home. Most biologists agree: Any high-school student can do it.

Many concerned scientists believe that recombinant DNA may well be the single most dangerous technology to come along in history. Dr. Liebe Cavalieri of the Sloan-Kettering Institute feels that the A-bomb, nerve gas, biological warfare, and the destruction of the stratospheric ozone layer by fluorocarbons are all less of a threat to human existence than recombinant DNA. "All of these dangers can, in theory if not in practice, be limited or controlled," notes Cavalieri. "The threat of a new life form is more compelling, for once released, it cannot be controlled and its effects cannot be reversed. A new disease may simply have to run its course, attacking millions in its path."[64]

According to Dr. Robert Sinsheimer of Cal Tech, "what we are doing is almost certainly irreversible. Knowing human frailty, these structures will escape, and there is no way to recapture them."[65]

Dr. Erwin Chargaff, the iconoclastic professor emeritus of Columbia University, is typically to the point when he prophesies, "I should say that the spreading of experimental cancer may be confidently expected."[66]

The possibility that these scenarios of mass death and genetic destruction could become reality was frightening enough to cause even the technology's major proponents to stop and think. On July 26, 1974, eleven of them took the unprecedented step of publishing an open letter to the biological community, calling upon their fellow researchers to observe a temporary moratorium on performing certain of the most dangerous experiments. The signers of the document read like a "Who's Who" of molecular biology. Among them were James D. Watson, Paul Berg, and Cohen and Boyer. With tne weight of this all-star cast of scientists behind it, the moratorium was almost universally observed.[67]

Predictably, the recombinant DNA moratorium—the first self-imposed ban on basic research in the history of science—provoked a storm of press interest. Scientists who had never gotten closer to a reporter than watching Walter Cronkite on television suddenly found themselves barraged by journalists wanting to know just what exactly was going on in their laboratories. Articles began appearing in national newspapers under headlines like "A Danger in 'Man-Made' Bacteria,"

"Hybrid Molecules Test Threat Seen," and "Bid to Ban Test Tube Super Germ."[68] Recombinant DNA, it was apparent, had become a subject of national interest.

There was one vital component left out of the early recombinant DNA debate, however. The safety issue was of such interest that no one, it seemed, was concerned about the application of the research to human genetic engineering. In reporting the recombinant DNA story, the press rarely delved into how the products of the research might one day be used. This left many people with the opinion that a few mad scientists were fooling around with dangerous chemicals for want of anything better to do with their time.

The researchers themselves, having brought unprecedented attention to their work, avoided any discussion of the application of the technique beyond a laundry list of possibilities that included new grains, antipollutant microorganisms, and that ultimate justification of any medical research, a cure for cancer. "I do not see this as a great moral or ethical issue," Paul Berg allowed. "I see this more as a problem of public health."[69] Most other molecular biologists were only too eager to agree.

SCIENCE POLICES ITSELF

Asilomar, California, is a retreat located in Pacific Grove, an idyllic resort town on the Pacific known principally for its white sand beaches, the annual return of the orange-and-black monarch butterflies, and its elegant golf courses. It was here, for three and a half days in February, 1975, that 140 biologists from 17 countries met to consider the fate of recombinant DNA.

For eight months these scientists had observed their self-imposed moratorium on certain experiments; from the opening moments at the Asilomar Conference, it was clear that they were eager to get on with business. Looking back on the conference, Dr. Erwin Chargaff finds it both lamentable and rather comical. "This was probably the first time in history," he observes, "that the incendiaries formed their own fire brigade."[70]

There were no social scientists present. Nor were there any ethicists, environmentalists, or theologians. None of the prominent scientific critics of the research was in attendance. There were no representatives of the group that was most at risk in the laboratories where the experiments were conducted: the lab assistants and technicians. And with the exception of a handful of reporters, largely representing scientific journals, there were no members of the constituency that was supposed to benefit most from the development of recombinant DNA technology: the public at large.

The conference had only one objective: to determine a set of rules and safety precautions under which recombinant DNA research could be resumed. Testifying before Congress several months after Asilomar, one attendee, Dr. Donald Brown of the Carnegie Institute, admitted that "there was no discussion of the applications of this research and the potential danger of these applications socially. There was no time for it. It was a very limited discussion."[71]

In many ways, the Asilomar Conference was analogous to the FBI's meeting to decide on how to regulate itself. Or, as one observer remarked afterward, having these particular scientists determine how to conduct their own research was like "having the chairman of General Motors write the specification for safety belts."[72]

The first three days of the conference were largely devoted to a parade of scientists who showed slides detailing their work for the benefit of their colleagues. Whenever the issue of safety guidelines was brought up, there was a clear indication that many, if not most, of the 140 attendees questioned whether there should be *any* regulations at all. As reported later in *Science News,* Asilomar was generally the scene of "unyielding, self-indulgent and conflicting attitudes."[73] It began to appear that by conference end the moratorium would be lifted and everyone would go home to their labs to conduct whatever experiments they wanted.

Then, on the evening of the third day, a note of reality intruded itself. Three lawyers gave presentations on the legal responsibility of researchers who create a "biohazard." The final speaker, Professor Harold Green of the George Washington University Law School, made the most telling point of the entire conference, as it turned out. His topic, he told the group, would be "Conventional aspects of the law, and how they may sneak up on you—in the form, say, of a multi-million-dollar lawsuit."[74] That hit home.

The next morning, at the final session of the conference, the assembled scientists endorsed a two-point safety program. First, they established broad, general categories of experiments that represented differing degrees of risk, and designated the type of laboratory precautions that would provide "physical containment" of potential hazards. Second, they developed the novel concept of "biological containment" by agreeing to use an enfeebled *E. coli* as a host for their recombinant DNA chimeras. The use of this biologically weak host, scientists hope, will prevent any chimera that might escape from a laboratory from living in the natural environment.

Subsequently, the broad, graded risks of Asilomar were redefined by the National Institutes of Health and drafted into detailed Guidelines that establish four levels of risk (P1–P4) and the laboratory requirements to meet them. These Guidelines are in effect today. As for the enfeebled *E. coli,* Dr. Roy Curtiss of the University of Alabama has developed a weak, but not foolproof, specimen which he has patriotically named "x-1776." It is now in wide use.

THE DEBATE OVER RECOMBINANT DNA

The moratorium announcement, the Asilomar Conference, and the NIH Guidelines were all unique attempts by the scientific community to police itself. Ultimately, however, they did more to provoke the controversy than contain it. Today, the recombinant DNA debate is more widespread than ever, much of it generated by molecular biologists who believe that their colleagues have given insufficient attention to the safety issues, and none at all to the social implications of the technology.

Critics contend that the possibility of a biohazard has not been resolved, just covered up. In the entire field of biological research during the past thirty years, at least 5,000 laboratory-acquired infections have been reported, one third of them in labs with special containment facilities.[75] Even at the home of America's biological-warfare research, Fort Detrick, the facility with the strictest containment procedures in the country, 435 laboratory-related infections have been documented.[76] "As long as people go into and out of the facility, which they must," says Dr. Jonathan King of MIT, "the bacteria will travel out, since it inhabits humans."[77] Once such a recombinant DNA type of bacteria escapes, there will be no recalling it.

There is also the question of anticipating the unknown. Since no one can realistically predict what they are creating with recombinant DNA, how can they establish safety guidelines to protect us from it, asks

Dr. Ruth Hubbard of Harvard. In the face of novel organisms, she argues, the current safety regulations "are largely an exercise in Newspeak, designed to contain anxieties about the hazard. Like the A-bomb shelter program of the early sixties, they promise protection against unpredictable and unquantifiable risks."[78] Dr. Willard Gaylin, testifying on recombinant DNA before a Senate committee, echoed the same theme. "It is of the nature of disasters," he warned, "that they not be anticipated, otherwise they would have been prevented."[79] The most serious disasters of science are therefore always the ones that remain unanticipated.

Voicing another concern, Dr. Robert Sinsheimer suggests that one of the unanticipated dangers in recombinant DNA is inherent to the technique, no matter what type of novel organism is created. He fears that the very act of artificially combining DNA from prokaryotes (or lower organisms, like bacteria) with that of eukaryotes (animals and plants) represents a fundamental and potentially dangerous violation of the thrust of three billion years of evolution. There is no proof at all, he contends, that such leaps backward across the genetic barrier separating lower and higher organisms take place naturally. What, he wonders, will be the results of this "introduction of quantum jumps into the evolutionary process?"[80] To that question, there is only silence. Again, as with so much of this technology, no one knows.

Still, there are many molecular biologists who utilize recombinant DNA technology who will argue that, in the words of Dr. Donald Brown, "the risks are nonexistent and the rules super-strict."[81] Several scientists, including Dr. Charles Thomas of the Harvard Medical School, have publicly stated that they would drink recombinant DNA if it would convince the public that there is no health hazard. (To date, none has followed through.) Joshua Lederberg is another scientist who believes that there is little if any danger involved in the use of recombinant DNA. The greater danger, he says, is establishing excessive safety precautions that create an "unjustifiable fear" among the populace. Conceding that eventually something destructive might escape from a lab, Lederberg resorts to a kind of macabre humor, calling what might result an "ecological upset" no more severe than "the crab grass in your lawn."[82]

Pursuing another course of argument, recombinant DNA researcher Dr. Arthur Kornberg darkly warns that if the use of the technique is curbed in this country, experimentation will proceed abroad, very possibly in "Moscow."[83] Dr. Norton Zinder, a former student of Lederberg's, questions the motives of the critics of recombinant DNA, calling them "ideologues emotionally against the idea of ever using genetic technology."[84] James Watson hints that if the public attempts to control recombinant DNA research it will go underground: "there should be no illusion that regulation is possible."[85]

Other proponents of the new technology take the broadest approach to the issue, placing the fate of the nation, and all of science, on the outcome of the current controversy. Curbing recombinant DNA research, argues Dr. Donald Frederickson of the NIH, will have "important impacts on American science, both in research and in the development of technology. The leadership of the United States in biological research would be threatened."[86] That possibility, says Paul Berg, would be "far too great a price to pay for fears that are so intangible, remote and imponderable."[87] Dr. Maxine Singer of NIH goes even further, placing the current debate in the context of a metaphysical battle between the forces of light and dark. Branding unnamed critics of the research as dissident elements embodying "uninformed fear, mysticism and political opportunism," Singer declares that the work "must continue; nothing less than science itself is at stake."[88]

Still, Dr. Robert Sinsheimer remains unconvinced.

"I can state my objective very simply," he writes. "The atomic age began with Hiroshima. After that, no one needed to be convinced that we had a problem. We are now entering the Genetic Age; I hope we do not need a similar demonstration."[89]

A major biohazard caused by the laboratory escape of a lethal recombinant DNA chimera would indeed be as dramatic a demonstration of the awesome power of the Genetic Age as Hiroshima was of the Nuclear. But barring an epic plaguelike disaster, history may well record that recombinant DNA's "revolutionary" potential was finally realized by its role as a prelude to systematic, long-range, human genetic engineering.

The use of recombinant DNA on human beings as a tool of genetic surgery is no more than a few short years away. In fact, so rapidly are new breakthroughs in the field piling up that the preceding sentence may be outdated by the time this book is published.

Sinsheimer, along with many other concerned biologists, believes the application of genetic engineering through recombinant DNA poses social hazards of the highest magnitude. "The potential for misuse is inherent in any scientific advance," he says; "it would seem particularly virulent here."[90]

The step-by-step process leading to the full-scale engineering of human development can already be anticipated. As an initial step, gene surgeons will use recombinant DNA to introduce new genetic material into a human to correct some simple deficiency or disorder—diabetes, for example. Hundreds of other diseases will also be "cured" by genetic intervention. Eventually, our new knowledge of DNA will be used to redirect more complex traits. Already, genetic engineers are exploring the chemical bases of schizophrenia and other psychological "problems." Finally, true genetic engineering will develop. Certain social groups may receive heightened intelligence or superior health courtesy of the bioengineers. Others, less powerful, may find their fate lies in genetic scrambling that destroys anger, frustration, resistance, and other modes of "antisocial" thinking. All of this is but a logical outcome of our current genetic program that calls for increased control of our biological processes.

Outside of the imagination of the researcher, there are virtually no limits to the possibilities raised by recombinant DNA and other biological methods of remaking Homo sapiens. In the fall of 1976, Dr. Roger Lewin, writing in *New Scientist* magazine, reported on an international conference of scientists engaged in genetic engineering. During the course of the meeting, he wrote,

> practitioners displayed an impressive array of tools with which to manipulate genetic machinery of all shapes and sizes, an array that leads one to suppose that even the most formidable technical barrier will be breached, given enough time. Genetic engineering has been going through an essential tooling up phase; it is now ready to go into business.[91]

At the close of the conference, Dr. Hans Kornberg provided a fitting testimonial to the brave new future: "we have described the journeys we are about to make rather than those we have already made."[92]

In fact, Kornberg's remark is only partially true. While the genetic engineers are being equipped for their journey with a caravan of scientific and technological know-how that could only be dreamed of just a few years ago, their final destination is one which many scientists have attempted to reach before, without success.

Bioengineering a better, or even "perfect," human specimen has long been the goal of geneticists. In this regard, genetics, unlike the other branches of science, has always had an ideological wing, which has been as indispensable to its development as the scientific discoveries themselves. The concept of perfecting the human race (or parts of it) through biological redesign has spawned some of the most controversial and brutal social and political movements of the twentieth century. Whenever dramatic new discoveries in genetics have been made, they have been translated almost immediately into political and social terms. Conversely, whenever the body politic (or parts of it) have become consumed with the desire to biologically modify or alter the social, economic, and political life of specific classes or groups of people, it has looked to and encouraged geneticists to experiment with new ways to accomplish such ends. This symbiotic relationship between genetic engineering and social policies and visions, each feeding off the other, each unable to exist without the other, found its high-water mark in the genetic policies of Hitler's Third Reich between 1932 and 1945.

What is less well known is that long before Hitler's rise to power, American geneticists and social ideologues had begun working closely together to fashion similar technologies and programs designed to both eliminate the so-called inferior stock from the human species and at the same time create a master race right here in this country. This close relationship between genetics and social dogma has existed for over seventy years, from the turn of the century to today. The only difference between today and yesterday is that in the past the tools of genetic modification were crude; now, because of the knowledge resulting from the Biological Revolution, they are as sophisticated as any technology on earth. On the other hand, the social ideology surrounding this research has not changed at all.

To appreciate the profound implications of the Genetic Revolution, it is essential to understand the history of the interrelationship between the science itself and the social philosophy that underlies it.

2

eugenics: the ideology behind genetics research

A MOVEMENT WRITTEN OUT OF AMERICAN HISTORY

> Some day we will realize that the prime duty, the inescapable duty of the good citizens of the right type is to leave his or her blood behind him in the world; and that we have no business to permit the perpetuation of citizens of the wrong type. The great problem of civilization is to secure a relative increase of the valuable as compared with the less valuable or noxious elements in the population. . . . The problem cannot be met unless we give full consideration to the immense influence of heredity[1] . . . I wish very much that the wrong people could be prevented entirely from breeding; and when the evil nature of these people is sufficiently flagrant, this should be done. Criminals should be sterilized and feebleminded persons forbidden to leave offspring behind them . . . The emphasis should be laid on getting desirable people to breed.[2]

This quote could have come from the lips of countless political functionaries at party rallies and meetings throughout Nazi Germany in the 1930s. But it didn't. It was uttered by the twenty-sixth President of the United States, Theodore Roosevelt, and it represented the "enlightened" view of millions of Americans caught up in an ideological movement that has been virtually written out of American history books. The movement was called eugenics, and in a thirty-year span (between 1900 and 1929) it captured the attention of America's leading reformers, academics, professionals, and political leaders. Had the movement succeeded, we might well have committed the very kind of atrocities that we tried the Germans for at the Nuremberg Trials in 1946.

Eugenics is generally categorized into two types, negative and positive. Negative eugenics involves the systematic elimination of so-called biologically unfit types of people from the population. Positive eugenics is concerned with the increased breeding of the most "fit" biologically sound types of persons in society.

From the turn of the century until the Great Depression, eugenics was embraced by the *crème de la crème* of the progressive forces as the cure-all for the economic inequities and social ills that were threatening the fabric of American life. It took hold at a time when reformers were becoming increasingly disheartened by their seeming inability to deal effectively with the escalating problems of crime, poverty, and social unrest.

The eugenics phenomenon was spawned in the wake of the first massive immigration wave, militant union-organizing drives, and the mushrooming growth of city slums in the late 1890s. It reached its peak in the chilling isolationist atmosphere following World War I, which produced the first great red scare in America. During this time, America's old-line ruling families combined forces with upper-middle-class academics and professionals in an active alliance to promote the notion of a eugenics policy for the U.S. For their part, the blue bloods—the Harrimans, Lodges, Saltonstalls, etc.—were becoming increasingly paranoid over their loss of control over the economic and political machinery of the country. For the first time, WASP hegemony was being vigorously challenged by the Irish, the Jews, the Italians, and other immigrant groups demanding a piece of the action in the early years of this century.

At the same time, the middle-class professionals and

academics were desperately looking for a way to explain away their failures in the area of social and economic reform. Both groups found the answer in eugenics. Its attractiveness was irresistible. First, its premise—that heredity, not environment, determined the action of people in society—allowed the reformers the excuse they needed to place the blame on the masses for the wrongs that beset society. The upper-class WASPs saw in eugenics a philosophical rationale that they could seize on in order to protect their "legitimate" claims to power. Most important, at a time when science was being heralded as the linchpin of American greatness and a road map to its manifest destiny, eugenics apparently offered a scientific explanation for social and economic problems and a scientific approach to their solution. In the words of Mark H. Haller, a historian and scholar in eugenics history, the eugenics movement became enormously powerful and influential precisely because it appealed to the "best people."

It was these "best people" who overnight turned eugenics into a form of secular evangelism. They preached their new-found creed in university lecture halls, before professional conventions, and on political platforms from one end of the country to the other. And the message was always the same: America's salvation hinged on its resolve to eliminate the biologically inferior types and breed the perfect human race.

To understand how such an alien ideology as eugenics could so captivate an entire generation of America's intellectual, professional, political, and financial elite, it is necessary to take a brief look at its beginnings and then sketch out its development and eventual application in American life in the first quarter of this century.

THE EARLY HISTORY OF EUGENICS

Eugenics is not a new idea. Writing in *The Republic,* Plato asserted that "the best of both sexes ought to be brought together as often as possible and the worst as seldom as possible and that the former unions ought to be reared and that of the latter abandoned, if the flock is to attain to first-rate excellence."

Caesar was so interested in improving the stock of Rome's best family lines that he offered a thousand sesterces to every "Roman" mother for each child. Augustus later offered two thousand, but to no avail. The birthrate among the rich continued to decline.[3]

The recognized father of modern-day eugenics was Sir Francis Galton. A cousin of Charles Darwin's, Galton was very much influenced by the publication of *Origin of Species.* Galton believed that Darwin's theory of evolution and the survival of the fittest also applied to the human species, so he set out to construct a theory which interpreted human social actions and behavior on the basis of his biological origins. In his book *Hereditary Genius,* Galton laid down much of the theory that was to be later used by so-called Social Darwinists to rationalize the worse abuses of unrestrained capitalism and racism in America. Galton's initial interest in this whole matter of eugenics (a term which he originated) stemmed from his desire to better understand why his family tree was decorated with so many outstanding personages. Galton was a fascinating character, one of those rare breed of English gentlemen who could honestly claim the title of true Renaissance Man. According to his biographers, Galton could read at age two and a half. During his lifetime he contributed to fields ranging from photography to meteorology.

He is especially well regarded by Scotland Yard, because he invented the fingerprint as a means of identifying criminals.[4] It's no wonder then that he strongly believed in what are called "natural talents." What's more, according to his peers, he was as well endowed physically as he was mentally. Beatrice Webb, the noted Fabian socialist, described Galton's appearance this way: "That tall figure with the attitude of perfect physical and mental poise, the clean shaven face, the thin compressed mouth with its enigmatic smile, the long upper lip and firm chin and as if presiding over the whole personality of the man the prominent dark eye brows from beneath which gleamed with penetrating humor, contemplative eyes. Fascinating to me was Francis Galton's all-embracing, but apparently impersonal beneficence."[5]

With attributes like those, it's not hard to understand why Galton might come to believe in the superiority of certain biological types.

In *Hereditary Genius* Galton concludes that the modern European (of which he considered himself the best of the lot) possesses much greater natural ability than do those of "lower" races. He then speculates as to the potential of a eugenics program. "There is nothing either in the history of domestic animals or in that of evolution to make us doubt that a race of sane men may be formed, who shall be as much superior mentally and morally to the modern Europeans, as the modern European is to the lowest of the Negro races."[6] Galton sums up his hopes for humankind's future by asserting that just as it is easy "to obtain by careful selection a permanent breed of dogs or horses gifted with peculiar powers . . . so it would be quite practical to produce a highly gifted race of men" by similar means.[7]

One man who read Galton's thesis, and decided it was time to put the theory into practice, was utopian socialist John Humphrey Noyes, the founder of the Oneida Colony in New York State. "Every race horse, every straight backed bull, every premium pig," said Noyes, "tells us what we can do and what we must do for man."[8] In 1869, Noyes had fifty-three women and thirty-eight men sign a pledge to participate in an experiment to breed healthy perfectionists by "matching those most advanced in health and perfection." The women pledged that they would "become martyrs to science," and with that the first American experiments in eugenics began.[9]

From his theory of eugenics, Galton concocted a new view of charity, one that would be later taken up by American eugenics reformers in their campaign to purify the racial stock of the nation. Charity, said Galton, should "help the strong rather than the weak, and the man of tomorrow rather than the man of today; let knowledge and foresight control the blind emotions and impetuous instincts."[10] So convinced was Galton of the wisdom of applying his theory to human beings that he regretted that "there exists a sentiment for the most part quite unreasonable against the gradual extinction of an inferior race."[11]

Just a few short years after Galton constructed his theories, new discoveries in genetics provided just enough meager evidence (later proven erroneous) to construct the thinnest "scientific" rationales for a eugenics movement. And that was all that many people needed to hop on board the eugenics bandwagon in America.

First, in 1900, geneticists rediscovered the laws of Gregor Mendel, an Austrian monk who discovered the theory behind transmission and distribution of traits by simple genes in certain plants. Applying Mendel's laws, geneticists "could make predictions about the number and type of offspring to be expected from different types of matings."[12] Mendel's laws were quickly taken up in plant and animal breeding, and geneticists came to believe that they could be soundly applied to humans as well.

Second, geneticists became convinced that most if not all traits are determined by single genes acting independently. (Like Mendel's law, this was later proven incorrect as it applied to humans.) This gave scientists confidence in their ability to breed better human stock.[13]

Finally, the distinguished German scientist August Weisman "produced experimental evidence that characteristics acquired by an organism from environmental pressures could not be inherited by its descendants."[14] This was used by eugenicists as proof that environmental reforms would have absolutely no effect on improving the human condition. In the question of better environment versus heredity, the hereditarians now claimed the scientific ammunition they needed to promote a full-scale eugenics effort for the country. All that was necessary to insure a ready and accepting audience was the enthusiastic support of major American scientists. The nation was not to be disappointed.

Between 1900 and 1915, leading American geneticists were largely responsible for spearheading the early eugenics movement. According to Kenneth Ludmerer in his seminal work, *Genetics and American Society,* nearly half the geneticists in the country became involved in the eugenics movement in one way or another. The reason, which is more than apparent in their own speeches and writings, was their "alarm by what they considered to be a decline in the heredity quality of the American people."[15] Scientists became active in leadership roles in the eugenics cause in the hope "that they could help reverse the trend."[16] With almost religious vengeance, these scientists jumped right up onto the center stage of American politics and demonstrated for all that their crusading fervor was at least on a par with their scientific accomplishments.

Without so much as flinching a muscle, Michael F. Guyer of the University of Wisconsin boldly proclaimed that "all available data indicates that the fate of our civilization hangs on the issue."[17] The famed geneticist

Edward G. Conklin dispassionately observed "that although our human stock includes some of the most intelligent, moral and progressive people in the world, it includes a disproportionately large number of the worst human types."[18]

Not to be outdone, Professor H. S. Jenings of Johns Hopkins informed the American public that:

> The troubles of the world and the remedy of these troubles lie fundamentally in the diverse constitutions of human beings. Laws, customs, education, material surroundings are the creations of men and reflect their fundamental nature. To attempt to correct these things is merely to treat specific symptoms. To go to the root of the troubles, a better breed of men must be produced, one that shall not contain the inferior types. When a better breed has taken over the business of the world, laws, customs, education, material conditions will take care of themselves.[19]

In 1906 the American Breeders Association set up the first functioning Committee on Eugenics. Its stated purpose was to investigate and report on heredity in the human race and to emphasize the value of superior blood and the menace to society of inferior blood.[20]

The Committee's membership included such famous people as Luther Burbank, David Starr Jordan, and Charles Davenport.

Four years later, Davenport convinced Mrs. E. H. Harriman (wife of the famous industrialist) to purchase a tract of land at Cold Springs Harbor, New York, where he established the Eugenics Record Office. Davenport, who was director, and Harry H. Laughlin, the superintendent, were soon to become the dominant voices in the eugenics drive during this period. By the way, according to Davenport, Mrs. Harriman's enthusiasm for the program was due to "the fact that she was brought up among well-bred race horses [which] helped her to appreciate the importance of a project to study heredity and good breeding in Man.[21]

After 1910 eugenics societies sprang up in cities all over the country. Among the most influential were the Galton Society of New York and the Eugenic Education Societies of Chicago, St. Louis, Madison, Minnesota, Utah, San Francisco, and Battle Creek, Michigan.[22] In 1913 the Eugenics Association began, and in 1922 the Eugenics Committee of the United States (later the American Eugenics Society) was formed.[23]

By World War I eugenics was a favorite topic not only in the schools and on political forums, but at women's clubs, church meetings, professional gatherings, and in popular magazines of the day.

The urgency of eugenicists' appeals often bordered on near hysteria, as when the president of the University of Arizona warned that it is "an optimist indeed who can see in our trend toward race degeneracy . . . anything other than a plight in which the race must find its final destiny in trained imbecility."[24]

Some even began to call for a basic change in our form of government to accommodate a eugenics ideology. One was William McDougall, chairman of the psychology department of Harvard University. McDougall so feared that democracy would eventually result in the "lower breeds" outnumbering the "best stock" and overtaking the machinery of the state that he openly advocated a caste system for America, based on biological differences, in which political rights would depend on one's caste.[25]

Academics were so convinced of the wisdom and virtue of applied eugenics that many of them threw all scholarly caution to the wind. "We know enough about eugenics," said Charles R. Van Hise, president of the University of Wisconsin, "so that if the knowledge were applied, the defective classes would disappear within a generation."[26]

By 1915, most of the leading educators already agreed with Irving Fisher, the well-known Yale economist, that "eugenics is incomparably the greatest concern of the human race."[27] It's not surprising, then, that by 1928 over three fourths of all the colleges and universities in America were teaching eugenics courses.[28] Their teachers were men like Earnest A. Hooton of Harvard, who preached that "crime is the resultant of the impact of environment upon low grade human organisms." "The solution to the crime problem," he told Harvard undergraduates, is the "extirpation of the physically, mentally and morally unfit or (if that seems too harsh) their complete segregation in a socially aseptic environment."[29]

The eugenics creed also found willing adherents within the media. The *New York Times* helped fan the eugenics hysteria with statements like "labor disturbances are brought about by foreigners" and "demonstrations are always mobs composed of foreign scum, beer smelling Germans, ignorant Bohemians, uncouth Poles and wild-eyed Russians."[30]

It might interest today's subscribers to the prestigious left-liberal magazines the *Nation* and the *New Republic* that the founders of both publications were crusaders for eugenics reform. Edwin Laurence Godkin, founder of the *Nation,* believed that only those of superior biological stock should run the affairs of the country,[31] and Herbert David Croly of the *New Republic* was convinced that blacks "were a race possessed of moral and intellectual qualities inferior to those of the white man."[32]

Imagine, if you will, a President of the U.S. writing in *Good Housekeeping* (a favorite forum of Presidents) that "there are racial considerations too grave to be brushed aside for any sentimental reasons." According to President Coolidge, biological laws tell us that certain divergent people will not mix or blend. Coolidge concludes that the Nordics propagate themselves successfully, "while with other races, the outcome shows deterioration on both sides."[33]

Even some of America's great heroes succumbed to the eugenics fever. Alexander Graham Bell was one of them. Speaking before the American Breeders Association in Washington in 1908, Bell remarked: "We have learned to apply the laws of heredity so as to modify and improve our breeds of domestic animals. Can the knowledge and experience so gained be available to man, so as to enable him to improve the species to which he himself belongs?" Bell believed that "students of genetics possess the knowledge . . . to improve the race" and that education of the public was necessary to gain acceptance for eugenics policies.[34]

Many modern-day feminists will be chagrined to learn that Margaret Sanger, a leader in the fight for birth-control programs, was a true believer in the biological superiority and inferiority of different groups. In some of the toughest-sounding words to ever come out of the eugenics movement, Sanger remarked that "It is a curious but neglected fact that the very types which in all kindness should be obliterated from the human stock, have been permitted to reproduce themselves and to perpetuate their group, succored by the policy of indiscriminate charity of warm hearts uncontrolled by cool heads." Sanger had her own solution to the problem of human biological contamination of society and better breeding: "There is only one reply to a request for a higher birth rate among the intelligent and that is to ask the government to first take the burden of the insane and feebleminded from your back . . . Sterilization," said Sanger, "is the solution."[35]

The eugenics ideology became so pervasive between 1900 and 1930 that some historians even attempted to rewrite world history from a eugenics perspective. Thus, David Starr Jordan, president of Stanford University, claimed that Rome's decline resulted from its frequent military conquests, in which its best blood was sent out to battle and scattered throughout the empire. This left Rome to the stable boys, slaves, and camp followers, whose poor biological stock multiplied and populated the city with an inferior genetic species.[36] Ironically, this kind of eugenics analysis of history led many like Jordan to become pacifists, on the grounds that war would take away and destroy the best blood of the nation.

It was this obsessive concern with the blood of the nation that so animated the eugenicists. Jordan best summed up the attitude of the supporters of eugenics when he declared that "the blood of the nation determines its history . . . the history of a nation determines its blood." To Jordan and his cohorts, "the survival of the fittest in the struggle for existence is the primal cause of race progress and race changes."[37] Just a few years later, a house painter from Munich was to echo that exact same sentiment from his jail cell in Germany as he put the final touches on a work which he entitled *Mein Kampf*.

Even the success of the fledgling Boy Scout movement in America was attributable to some degree to the interest in eugenics. As a matter of fact, David Starr Jordan served as the vice-president of the Boy Scouts of America in those early days. Jordan and his colleagues believed that the Scout program could help rear the "eugenic new man."[38]

One of the most bizarre twists in the eugenics movement was Fitter Family Contests, run by the American Eugenics Society. Blue ribbons were presented at county and state fairs all over the Midwest to those families that could produce the best family pedigrees. Families were judged on their physical and mental qualities, right alongside pigs and cows.[39]

The acceptance of eugenics by much of the general public as a scientifically sound theory was due in large part to the early and enthusiastic support of eugenics by some of America's most prominent scientists. The scientists legitimized the theory of eugenics in the public mind, although in the end they largely refused to accept any responsibility for the consequences that flowed from its application. The story of its ruthless application represents one of the darkest pages in American history.

THE FIRST BATTLEGROUND: STERILIZATION

Eugenicists looked on sterilization as a major tool in their campaign to weed out biologically inferior stock from the American population. As a result of a systematic and coordinated drive by eugenics advocates, tens of thousands of American citizens were involuntarily sterilized under various laws enacted by the individual states after the turn of the century.

Much of the early rationale for these sterilization laws was provided by experts in the field of criminology who borrowed liberally in their analysis from the geneticists. The first major work attempting to link criminality and other antisocial behavior with heredity was published in 1876 by famed criminal anthropologist Cesare Lombroso. Among other things, Lombroso argued in his study, *The Criminal Man,* that potential criminals (which included everyone from murderers to prostitutes, and thieves to paupers) "tended to have a primitive brain, an unusual cephalic index, long arms, prehensile feet, a scanty beard but hairy body, large incisors, flattened nose, furtive eyes and an angular skull. Like a savage the criminal often tattooed himself."[40]

Believing that criminals were born and not made, Lombroso claimed that it was even possible to draw a composite profile of those people likely to commit particular types of crime. For example, "thieves were characterized by small restless eyes, thick eyebrows, a crooked nose, thin beard and a narrow receding forehead." "Sex criminals had bright eyes, a cracked voice, blond hair and a delicate face." "Murderers had cold, glassy eyes, long ears, dark hair and canine teeth."[41] Ironically, Hollywood still uses Lombroso's stereotypes in casting for criminal roles.

By the 1890s medical experts and superintendents of state penal institutions were generally convinced that heredity and crime went hand in hand.

Henry M. Boies, an expert on penology of the Pennsylvania State Board of Public Charities, wrote in 1893 that it is "established beyond controversy that criminals and paupers both are degenerates: the imperfect, knotty, knurly, worm eaten fruit of the race."[42] By 1900 heredity was being blamed for behavior ranging from criminality to eccentricity and alcoholism, and W. Duncan McKinn, in his book *Heredity and Human Progress,* was arguing that "the surest, the simplest, the kindest, and most humane means for preventing reproduction among those whom we deem unworthy of this high privilege is a gentle and painless death."[43]

Sterilization as a eugenics cure for everything from crime to feeblemindedness began to come into vogue just before the turn of the century.

In the mid-1890s the superintendent of the Kansas State Home for the Feebleminded castrated forty-four boys. Even though public opinion forced him to stop his program, the Board of Trustees replied that "those who are now criticising Dr. Pilcher will, in a few years, be talking of erecting a monument to his memory."[44] It never happened.

Castration was not very popular with the general public, so when the first eugenic vasectomy was performed in 1899 it was heralded as a more efficient tool and was immediately embraced by the pro-sterilization forces. The vasectomy was performed at Indiana Reformatory by Dr. Harry Sharp as a eugenics solution for a nineteen-year-old who had become "addicted" to excessive masturbation. Within three weeks of the operation, according to Sharp, the lad was excelling in his classwork once again. So encouraged was Sharp by the results that he took it upon himself to operate on seventy-six more youths by the end of the year.[45]

The same state, Indiana, passed the first sterilization law, in 1907. The bill called for mandatory sterilization of confirmed criminals, idiots, imbeciles, and others in state institutions when approved by a board of experts. It was later fondly referred to by eugenicists as "the Indiana Idea."[46]

As to why there was such a willingness to believe in the hereditary basis of crime, dependency, and delinquency, one reformatory superintendent pointed out that the experts were being pressured by the public to explain the apparent inability of state institutions to rehabilitate. He observed that "the only way in which their criticism can be met is by producing data showing that a large majority of these failures were due to mental defect on the part of the inmates and not to faults in the system of training."[47] Once this line of thought was accepted, sterilization became the easiest and most logical solution to the problem.

It should be pointed out, in passing, that not all eugenics supporters were enthusiastic about sterilization. There were those who worried that such policies, if pursued on too large a scale, could have a deleterious effect on the morality of the country. Walter E. Fernald, a leading name among institutional superintendents, told the story of a feebleminded girl in a small town who had an ovariectomy. The men then began to stand in line on her front porch to have intercourse with her after her mother went off to work each morning. The result, said Fernald, was an epidemic of gonorrhea.[48]

Between 1907 and 1917 two new propaganda techniques were introduced by eugenics advocates, both of which greatly increased support for broader use of sterilization procedures.

First, eugenicists began to use family trees as a major propaganda weapon in their battle to equate heredity with inferior biological types. By concocting elaborate family histories, the eugenicists purported to show that inferior traits were passed down from generation to generation and that the social cost of allowing such reproduction to continue was not worthwhile.

The most famous of these family histories was that of the Kallikaks. Deborah Kallikak was a little moron girl under institutional care. The Eugenics Records Office decided to trace her family tree in an effort to prove that feeblemindedness was hereditary. According to Elizabeth Kite, the field researcher on this project, Deborah's great-great-great-grandfather Martin Kallikak, while fighting in the American Revolution, engaged in sexual intercourse with an unknown feebleminded girl at a tavern, which resulted in an illegitimate son. The son, nicknamed "Old Horror," had 480 descendants, of whom 143 were feebleminded, 46 normal, and the rest of doubtful or unknown mentality. The descendants included 26 illegitimate children, 33 sexually immoral persons, 24 alcoholics, three epileptics, and three criminals. The important thing about the Kallikak story, however, is that after the Revolution this same great-great-great-grandfather married a nice upstanding Quaker girl of good family, and from this marriage came a line of doctors, lawyers, judges, and educators—proof, said the eugenics advocates, that had they only sterilized that feebleminded colonial girl, the public would have been saved a great deal of tax money and anguish.[49] The Kallikaks were followed up by case histories of other families, such as the Jukes, the Tribe of Ismael, the Family of Sam Sixty, the Happy Hickory Family, the Hill Folk, the Nam Family, all traced by the Eugenics Records Office.[50] And, of course, with the publication of each successive family tree, the eugenicists claimed more proof that crime, prostitution, alcoholism, pauperism, and feeblemindedness were hereditary and that sterilization was the solution.

The second element that eugenicists introduced to push their claims was the Binet intelligence test. The tests were administered among the institutionalized to prove them mentally and biologically incompetent and a drain on society that should not be allowed to reproduce. Together the family trees and the Binet tests fueled the propaganda fires even further and hastened the passage of more state laws on sterilization.

In 1914, Harry H. Laughlin (himself childless) issued a report to the American Breeders Association in which he stated that "society must look upon germ plasm as belonging to society and not solely to the individual who carries it." The most astounding part of his report was the finding that 10 percent of the population of the U.S. were "socially inadequate biological varieties" who should be segregated from the general population and sterilized.[51] Among those Laughlin targeted for sterilization were epileptics. What makes this particularly interesting is the fact that Laughlin himself was an epileptic. There is no record, however, of his ever having been sterilized.

With the chorus of demands for sterilization reaching a crescendo, fifteen more states enacted laws between 1907 and World War I. The extent to which sterilization mania was carried is perhaps best reflected in a bill introduced in the Missouri legislature calling for sterilization of those "convicted of murder, rape, highway robbery, chicken stealing, bombing, or theft of automobiles."[52]

The constitutionality of these sterilization laws was not tested until 1927, when the Supreme Court ruled in a Virginia case that sterilization fell within the police powers of the state. What was even more incredible than the ruling was the opinion of the esteemed jurist Oliver Wendell Holmes, who wrote:

> We have seen more than once that the public wel-

> fare may call upon the best citizens for their lives. It would be strange if it could not call upon those who already sap the strength of the state for these lesser sacrifices, often felt to be such by those concerned, in order to prevent our being swamped with incompetence. It is better for all the world, if instead of waiting for their imbecility, society can prevent those who are manifestly unfit from continuing their kind . . . three generations of imbeciles are enough.[53]

By 1931, thirty states had passed sterilization laws and thousands upon thousands of American citizens had been surgically fixed.[54]

EUGENICS THEORIES DISPROVED

Even as the eugenics movement was gaining converts and momentum, new discoveries and insights in the social sciences and in biology were challenging the basic scientific "truths" that eugenicists were operating under. By 1914, the gap between scientific facts and eugenics fiction had become so great that some geneticists, psychologists, and social workers who had previously rushed onto the eugenics bandwagon were moderating their enthusiasm and their active participation.

First, the intelligence tests used by eugenicists to measure inferior biological types were literally invalidated by none other than the U.S. Army. In a somewhat humorous episode, the military tested over 1,726,000 recruits for intelligence during World War I and found that nearly half were rated as feebleminded. The implications were embarrassing: Either half the population of the U.S. was biologically unsound and mentally incompetent, and therefore proper subjects of sterilization and institutionalization—which the government was not prepared to acknowledge—or the testing techniques were pure bunkum. Even more embarrassing, according to the tests results, blacks in five Northern states scored higher than whites in eight Southern states.[55] The only conclusion that could be drawn was that either Northern blacks were genetically superior to Southern whites, or environment played the dominant role in determining intelligence. Obviously the eugenicists were quite unprepared to argue that Northern blacks were biologically superior to all the decent Southern whites.

Then there was the question of the methodology used by the eugenicists in tracing the genealogy of families like the Kallikaks. Some social scientists began to ask how the Eugenics Record Office could know that Deborah Kallikak's great-great-great-grandmother was feebleminded, since the report itself acknowledged that her name was not known and there were no records of any kind about her. A closer look at the Kallikak report and the others indicated that they were little more than the fictional observations of the field workers putting them together. For example, Kite's own account of her research into the Kallikak family is replete with "rigorous" scientific observations of the following kind: The father, too, "though strong and vigorous, showed by his face that he had only a child's mentality. The mother in her filth and rags was also a child."[56] So much for scientific eugenics.

Meanwhile, in the social sciences, more accurate mental testing, new insights into the psychology of early childhood training and its effects on character development, and new diagnostic techniques in social work demonstrating correlations between crime, economic condition, and environmental influences all hammered away at the pseudo-scientific assumptions of the eugenicists' creed.

Most important of all, a series of new genetics discoveries did much to destroy the major "scientific" principles employed by the eugenicist. A Danish botanist, Johansen, found that environment was the cause of more variation among different plant strains than their genetic make-up was. Two scientists, George H. Hardy and Wilhelm Weinberg, discovered that eliminating particular genetic traits from a population was a much longer and more difficult process than had been thought. And in 1913 Edward East and Rollins Emerson disproved the notion that most, if not all, traits are determined by single genes.[57]

It is important to emphasize, however, that despite these new insights and discoveries, most geneticists continued to maintain their interest in the eugenics movement, though at a much lower profile. In fact, in 1921 an assortment of some of the most prominent geneticists in the country presented reports and papers at the International Congress of Eugenics.[58] With only a few exceptions, the scientists did not step forward and publicly challenge the discredited "scientific" assumptions that the eugenicists were citing to build their case. For this reason, the effect of these new insights and discoveries was minimal, and the eugenics movement continued to grow and to win wider support among elected officials and the general public over the next fifteen years.

EUGENICS' FINEST HOUR

The eugenicists' greatest triumph came after World War I in their successful campaign to enact an immigration law based on eugenics standards. The law which was passed in 1924, and which remained on the books until 1965, had the effect of altering the entire ethnic and racial composition of the U.S. to satisfy the standards laid down by eugenics supporters.

Eugenicists early on saw the value of mounting an ambitious drive to restrict immigration along "biological" lines. Writing to Charles Davenport in 1912, Irving Fisher of Yale remarked: "Eugenics can never amount to anything practically until it has begun, as Galton wanted it, to be a popular movement with a certain amount of religious fervor in it and as . . . there is already a sentiment in favor of restricting immigration . . . this is a golden opportunity to get people in general to talk eugenics."[59]

The eugenicists already had a convenient launching pad for their campaign in the results of a Congressional commission set up in 1907 to investigate the impact of immigration on the country. Composed of three Representatives, three Senators, and three Presidential appointees, the distinguished Dillingham Commission boasted a staff of three hundred and a budget in excess of $1 million. The conclusion of its three-year study was that immigrants from Mediterranean regions were biologically inferior to other immigrants.[60] This unofficial government endorsement of their own claims was just what the eugenicists needed to begin their offensive.

The opportunity to mount an all-out public drive for restrictive immigration came after World War I. The new mood of nationalism and isolationism, renewed labor strife, fear of a Marxist takeover of the country in the wake of the Russian Revolution, and the hordes of immigrants streaming into Ellis Island all created an ideal atmosphere for eugenics legislation. Overnight, the eugenicists began cranking up a new series of studies and reports to demonstrate the biological inferiority of certain immigrant groups.

Liberal sociologist Edward A. Ross, after a sixteen-month study, published a report entitled *The Old World in the New,* in which he claimed among other things that the Mediterraneans were prone to sex and violence and were irrational by nature; the Slavs were a passive people imbued with ignorance and superstition, the men wife beaters and alcoholics; and the Jews were clannish, tricky, and underhanded in business.[61] Another eugenicist, Madison Grant, added to the list with the Hindus, "who have been for ages in contact with the highest civilizations, but have failed to benefit by such contact either physically, intellectually or morally," and the Negroes, who "are the Nordic's willing followers and who ask only to obey and to further the ideals and wishes of the master race."[62] The Nordics, on the other hand, were found to be a race of "great energy and industry, vigorous, imaginative and highly intelligent."[63]

The first legislative battle over new restrictive immigration laws came in 1921. In an attempt to halt the influx of Southern Europeans, Congress moved to limit immigration to 3 percent of the foreign-born of each nationality in the U.S. according to the census of 1910.[64] This emergency legislation was merely a temporary stopgap, however, with the major immigration battle coming three years later, in the 68th session of Congress.

Restricting immigration was a particularly difficult issue for the nation's leading industrialists as well as for organized labor. Caught in a catch-22, the industrialists were very much in favor of a continued supply of cheap immigrant labor, but worried that it would bring with it increased organizing demands and Marxist notions of proletarian revolution. These latter fears were enough to convince some of America's business leaders to support restrictive immigration. Labor also felt trapped. On the one hand, more immigrants meant more recruits to the labor movement. On the other, increases in immigration would further depress an already weakened employment market. The pocketbook won out, and labor decided to support restrictive legislation.[65]

The nerve center for the eugenics campaign for restrictive legislation was the House Committee on Immigration and Naturalization, headed by Albert Johnson. If there was ever any question as to where the committee stood on the matter of immigration, it was quickly put to rest with the appointment of Harry H. Laughlin as the "Expert Eugenics Agent" of the committee, a post which allowed him to control the course of the proceedings.

Laughlin told the committee that, "making all logical allowances for environmental conditions . . . the recent immigration as a whole, present a higher percentage of inborn socially inadequate qualities than do the older stocks."[66] Throughout the hearings, no one was called to refute Laughlin's findings.

The Secretary of Labor, James J. Davis, expressed the sentiments of the Coolidge administration on the question of restrictive immigration, and in so doing set the tone for the ensuing debate: "America has always prided itself upon having for its basic stock the so-called Nordic race . . . we should ban from our shores all races which are not naturalizable under the law of the land and all individuals of all races who are physically, mentally, morally and spiritually undesirable and who constitute a menace to our civilization."[67]

The House debate, if it can be called that, was little more than a testimonial to the extent to which the eugenics hysteria had taken over and engulfed the nation. Representative Robert Allen of West Virginia took to the floor to inform his colleagues that "the primary reason for the restriction of the alien stream is the necessity for purifying and keeping pure the blood of America."[68] Echoing Allen's beliefs, J. Will Taylor of Tennessee warned that "America is slipping and sinking as Rome did and from identical causes. Rome had faith in the melting pot as we have. It scorned the iron certainties of heredity, as we do. It lost its instinct for race preservation, as we have lost ours."[69]

Putting it in a more folksy perspective, Thomas W.

Phillips of Pennsylvania said: "We know better than to import vicious or refractory animals but, on the contrary, through intelligent and careful selection from abroad, bend every effort to improve our homestock of domestic animals . . . We must set up artificial means through legal machinery to hand pick our immigrants if we are to prevent rapid deterioration of our citizenship."[70]

And so it went, as Representative after Representative took to the floor to reaffirm the commitment to a "eugenic America." The opposition forces were small and disorganized, composed chiefly of Congressmen of Jewish, Italian, and other ethnic backgrounds. Their voices were all but drowned out amid the clamor to "purify the blood of the nation."

In the end, the legislation passed the House with only 35 Republicans and 36 Democrats voting against it. The President signed the new immigration legislation, which called for restrictions based on 2 percent of the foreign-born from each country according to the 1890 census. Since the number of Southern Europeans coming into the United States was not far greater than the number of Northern Europeans, this law had the effect of closing the door on these new ethnic groups.

Emanuel Cellar, a freshman in that 68th Congress, was one of the voices in opposition to the legislation. Being Jewish and representing a largely Jewish and otherwise ethnic district in New York, Celler was appalled by the attitudes expressed during that debate. Over forty years later he was a prime architect in getting the legislation changed. In the interim, the American government operated under a eugenics standard in its overall immigration policy.

From its peak in 1924, the eugenics movement steadily declined, eventually collapsing in on itself five years later. The crash of 1929 had a profound impact on eugenics philosophy. With America's financial elite jumping out of windows and middle-class professionals and academics standing in unemployment lines right alongside Italian, Polish, and Jewish immigrants, it was no longer possible to retain the myth that there was something biologically superior about certain kinds of people. The Depression served as the great equalizer as millions of Americans, Nordic and Italian, WASP and Jew, found themselves in the same circumstances of poverty and destitution. Biological distinction gave way to a shared sense of common plight for all those forced into the bread lines of the Depression.

H. J. Muller, the eminent geneticist, thrust the final dagger into the early eugenics movement in a dramatic speech delivered before the Eugenics Society in New York in 1932. Reflecting the growing resentment toward corporate financiers and the capitalist system, Muller attacked the inverted logic that eugenicists had used for years to rationalize the excesses and abuses perpetrated by America's corporate elite. Muller challenged the assumption that the well-to-do were biologically of better stock than the poor, and questioned a eugenics dogma based on maintaining these so-called better specimens. It would be wrong, he said, to foster a eugenics program now because it would be used to further enhance the position of those in power at the expense of the masses of people. In 1932, people were finally ready to listen to such arguments.

Hitler's rise to power in 1932 was the other major factor behind the decline in importance of the eugenics movement in America. Ironically, in 1925, H. H. Newman, professor emeritus of zoology at the University of Chicago, had written the following lines:

> One needs only to recall the days of the Spanish Inquisition or of the Salem witchcraft prosecution to realize what fearful blunders human judgement is capable of, but it is unlikely that the world will ever see another great religious inquisition or that in applying to man the newly found laws of

heredity there will ever be undertaken an equally deplorable eugenic inquisition.[71]

That same year, German officials were writing away to state governments for information on American sterilization laws. Boeters, the leading advocate of eugenics in Germany at that time, remarked that "What we racial hygienicists promote is not at all new or unheard of. In a cultural nation of the first order, the United States of America, that which we strive toward was introduced long ago. It is all so clear and simple."[72]

While Boeters and others were busy reading reports about sterilization laws in the U.S. that year, the first printed edition of *Mein Kampf* was published in Germany. In it Hitler proclaimed: "The mixing of higher and lower races is clearly against the intent of nature and involves the extinction of the higher Aryan race . . . Wherever Aryan blood has mixed with that of lower peoples the result has been the end of the bearers of culture."[73]

Hitler's Third Reich came to power in 1932. Almost immediately the Minister of Interior, Wilhelm Frich, announced to the world that "the fate of racial hygiene of the Third Reich and the German people will be indissolvably bound together."[74] On July 14, 1933, the Führer decreed the Heredity Health Law, a eugenics sterilization statute which was to be the first step in a mass eugenics program that would claim the lives of millions upon millions of people over the next twelve years. In response to Hitler's unfolding eugenics campaign, American eugenicists observed that Germany was "proceeding toward a policy that will accord with the best of thought of eugenicists in all civilized countries."[75]

As a final chapter to the role of American scientists in the eugenics movement, throughout the 1930s the Genetics Society of America debated over and over again at its annual meetings the question of whether to formally condemn the eugenics policies of the Third Reich. There were never enough votes to approve such a condemnation.[76]

In 1936, the University of Heidelberg awarded an honorary degree to Harry Laughlin for his great contribution to the field of eugenics.[77]

THE REEMERGENCE OF THE EUGENICS MOVEMENT

The American eugenics movement lay virtually dormant during World War II. Then, in 1945, the U.S. dropped the first atomic bomb on Hiroshima. Several years earlier H. J. Muller had discovered that radiation from X-rays increases the mutation rate of genes. Scientists like Muller, and government officials at the Atomic Energy Commission, realized that the Atomic Age would bring with it a dramatic increase in radiation, which in turn would have serious and harmful effects on human gene mutation. The grave concern over atomic radiation and gene mutation sparked a renewed interest in genetics research.

The Atomic Energy Commission began to pour funds into genetics research in an effort to better understand the effects of radiation on human genes. One of their grants came in the form of a pre-doctoral fellowship to a young geneticist at the University of Indiana named James Watson. He used those funds to unravel, for the first time, the molecular structure of DNA.[78] Watson won a Nobel Prize for his brilliant discovery and opened up a new era in genetics research.

Muller's and Watson's pioneering work triggered a renewed interest in eugenics as well. Many prominent scientists began to claim that increased radiation was causing massive mutations in the human gene pool, which in turn were spreading with each new generation of babies. Likewise, it was argued that new breakthroughs in medical treatment of genetically defective individuals were keeping alive more biologically unfit people, who in turn were passing on gene defects to their offspring. If left to spread, scientists like Dr. Leroy Augenstein argued, the genetic load (defective genes in the population) would become so great "that within five to ten generations one in ten children will be seriously defective."[79]

As a result of these concerns, a whole new generation of modern-day eugenicists has emerged from within the scientific community. They are advocating a range of new eugenics procedures and programs to both arrest the "deteriorating trend" in the human gene pool and upgrade the genetic worth of the human species.

Bernard Davis, professor of bacterial physiology at Harvard, speaks for many of his scientific colleagues when he says that we need a eugenics program "aimed primarily at reducing the production of individuals whose genetic endowment would limit their ability to cope with a technologically complex environment."[80]

While such pronouncements are troubling in and of themselves, they become of even greater concern when prominent scientists attempt (as the eugenicists did at the turn of the century) to convince the public that reforms of the environment are useless because genotype is the primal cause of one's actions in society. Several years ago Sir Julian Huxley, the eminent biologist (and Aldous's brother), made just such a claim:

> It is clear that for any major advance in national and international efficiency we can not depend on haphazard tinkering with social and political symptoms or ad hoc patching up of the world's political machinery, or even on improving education, but

> must rely increasingly on raising the genetic level of man's intellectual and practical abilities.[81]

Huxley was not speaking in a vacuum. Once again scientists in universities all over America are beginning to issue reports and studies that purportedly demonstrate that crime, poverty, and intelligence are more the direct result of people's genetic make-up than of the environment in which they live. In fact, a whole new branch of science called sociobiology has emerged; its basic supposition is that social actions like aggression, cooperation, and violence are biologically determined. Exactly how new is this school of thought? Interestingly enough, in October, 1972, the old American Eugenics Society voted unanimously to change the organization's name—after almost sixty years—to the Society for the Study of Social Biology.[82]

This new mood of biological determinism comes in the wake of great political, social, and economic upheavals over the past two decades. The inability of governmental and economic institutions to deal effectively with the problems of rising crime rates, mass unemployment, urban decay, pollution, and a host of other critical problems is fostering a deep sense of cynicism and despair within the general public. Like Huxley, many scientists are now saying that the problem lies not in the institutions, but in the genetic makeup of the people. Nowhere is this trend toward biological determinism more evident than in the new research studies into criminal behavior.

One such research study was conducted over a six-year period at St. Elizabeth's Hospital in Washington by Dr. Sam Yochelson and Dr. Stanton Samenow. Over this extended period Yochelson and Samenow interviewed hundreds of convicts who had been placed at St. Elizabeth's because the courts had decided they were "disturbed people." According to the St. Elizabeth's study, criminals are not made by society, they are born that way. Samenow says that their experiments had resulted in "identifying 52 thinking patterns that are present in all criminals." "In a sense," says Samenow, "you would say he [the criminal] is programmed for killing, not by someone else, but he has a mind that reacts to adversity, not with surmounting adversity, but eliminating it."[83] Samenow's conclusion is that the solution "isn't just changing his [the criminal's] environment." "We found," says Samenow, "that the task of change involves changing a man's thinking."[84]

Drs. Samenow and Yochelson are not alone in their views on the genetic basis for criminality. Several years ago the federal government's Law Enforcement Assistance Administration (LEAA) gave a grant of $109,000 to Vernon Mark, director of neurological services, Boston City Hospital, Frank Erwin, UCLA Medical School, and William Sweet, chief of neural surgical services at Massachusetts General Hospital, to "develop a technique by which 'violence prone' individuals could be identified." The assumption upon which the grant was awarded is that criminal behavior is, in fact, a biologically identifiable trait. The researchers set forth their thesis that criminal behavior is genetic, not environmental, in a series of rhetorical questions: "If slum conditions alone determined and initiated riots, why are the vast majority of slum dwellers able to resist the temptations of unrestrained violence? Is there something peculiar about the violent slum-dweller that differentiates him from his peaceful neighbor?"[85] They recommend further studies "to pinpoint, diagnose and treat those people with low violence thresholds before they contribute to future tragedies."[86]

The chairman of the Department of Psychiatry at UCLA was also so convinced of the biological basis of certain kinds of criminal behavior that in 1972 he called for the creation of a Center for the Prevention of Violence. Its objective could be to "identify certain individuals who are characterized [biologically] by a very high probability of committing individual acts of violence." The funds were to come from the State of California via the California Council for Criminal Justice—a branch of LEAA.

In an address before a conference sponsored by the National Institute of Mental Health, Dr. O. J. Andy, chairman of the Department of Neurosurgery at the University of Mississippi, went so far as to claim that "all abnormal behavior results from structurally abnormal brain tissue." Rather than waste time and money on psychiatry and social-reorientation programs, which Andy contends are futile in many cases, surgical treatment, he believes, is preferable. Andy has performed dozens of such neurosurgical procedures already.

At the turn of the century, criminal anthropologists were measuring head sizes to find out who the criminal types were; today some scientists are looking at chromosome combinations. One recent controversy over chromosomes and criminal behavior is illustrative of the trend and the unwarranted assumptions that underlie it.

Over the last several years, garish stories have appeared in the daily press about a rare breed of "supermales" who have been programmed for murder and other crimes because of an extra chromosome in their genotypes. The normal male is born with an XY pair of sex chromosomes. In rare cases, a male will be born with an extra chromosome from his male parent. All of a sudden, several years back, some scientists began arguing that this special XYY individual was prone to aggressive antisocial and criminal behavior because of his super-maleness. Public interest in this theory was sparked when the press reported that mass murderer Richard Speck was an XYY type. This later turned out to be a false rumor, but it didn't stop scientists from writing over two hundred reports in specialized scientific journals on the subject. Writing in *Science* magazine, Bentley Glass, former president of the American Association for the Advancement of Science, even went so far as to say that he "looks forward to the day when a combination of amniocentesis and abortion will rid us of . . . sex deviates such as the XYY type."[87] Massachusetts and Maryland actually screened male adolescents in juvenile homes for the extra Y chromosome, and genetic-counseling services for pregnant women did the same in Denver, New Haven, and Boston for a time.[88]

Yet in the only comprehensive report done on the subject, D. S. Borgoankar and S. A. Shah found that the chances of abnormal behavior manifesting itself in an XYY male are no greater than in an XY male.[89] It appears that the only significant difference in the XYY type that has been found is a tendency toward greater height.

Again, like the eugenicists of sixty years ago, some scientists today are attempting to show that intelligence is a function of genes and not environment. In 1969, A. R. Jensen published an article in the *Harvard Educational Review* entitled "How Much Can We Boost I.Q. and Scholastic Achievement?" Jensen asserted that differences in intelligence are more directly correlated to inheritance than to environment. In fact, he boldly claimed that 80 percent of all IQ performance is inheritable. Jensen went on to contend that differences between blacks and whites, and between working-class and middle-class people, are chiefly the result of their genetic inheritance and therefore are unchangeable by environmental reform alone. For example, on the question of race and social performance Jensen concludes that "genetic factors are strongly implicated in the average Negro-white intelligence difference. The preponderance of evidence is, in my opinion, less consistent with a strictly environmental hypothesis than with a genetic hypothesis . . ."[90]

Since Jensen's original paper, many prominent scientists have made similar claims. Such thinking has even begun to influence some of America's most influential educators. Dean Watkins, chairman of the Board of Regents of the University of California, the largest system of higher education in the country, is one of them. "It is just possible," says Watkins, "that the reason some people are rich is because they are smarter than other people; and maybe they produce smarter children."[91]

Despite the fact that the theories of Jensen and many of his colleagues have been discredited in numerous studies, they continue to gain greater acceptance among academics and professionals across the country.

Support of the heredity-intelligence hypothesis has spawned a myriad of new eugenics proposals by some of America's "best people." Charles Frankel, professor of Philosophy and Public Affairs at Columbia, warns:

> The day may not be that far away when genetic manipulation to improve intelligence will be a cheaper and more practical means to get results than environmental reforms which have already proven to be more difficult to bring about and more disappointing in their consequences than was supposed. There are many mounting signs that the eugenicist's dream of a remade breed, so long in disgrace in consequence of Nazi forays into the field, may be on its way to a comeback.[92]

Scientist Richard Hernstein is less tentative about the future. He recommends that the government begin now to gather IQ data as part of the national census so that "we could observe dysgenic or eugenic trends in American society." Hernstein says that this approach makes good practical sense because "if at some time in the future we decide that our population is getting too large and we need to limit it, we could use census information on IQ to decide how and when to limit it."[93]

William Shockley, a Nobel laureate, goes even further. Worried that the intelligence level of the country is falling because blacks and white working-class people ("whose inherited intelligence is allegedly inferior") are outproducing the white middle class, Shockley advocates a program of cash rewards for sterilization linked by a sliding scale to the sterilizee's IQ score.[94]

If this all seems too farfetched for practical consideration by the federal government, think again. The National Institutes of Health has already funded a $1.7-million study at the University of Hawaii's Behavioral Biology lab to test 3,200 families of Caucasian and Japanese heritage. In an early draft of the grant application the scientists stated that "With respect to long-range significance these data will serve as a basis for future decisions about the disturbing but inevitable questions about population control which will have to be made at every level . . . the purpose of this study is to provide some solid information about the genetic correlates of intelligence so that an informed decision may eventually be made."[95]

The beginnings of a new eugenics movement are well under way in America. Forty-six states already have some type of mandatory genetic screening program for newborn infants,[96] and many scientists, reformers, and public officials are advocating that they be upgraded to cover a more extensive range of negative and positive eugenics programs. A recent screening law drawn up by the Chicago Bar Association and presented to the Illinois State Legislature is indicative of the eugenics trend: It would require genetic screening of all applicants for marriage licenses. A supporter of the bill voiced one of the underlying objectives of the proposed statute when he said that "we are going to have to try to reduce the number of non-productive members of our society."[97]

Some of America's best legal minds are already preparing background papers for possible eugenics legislation on the national level. William T. Vukowich, associate professor of law at Georgetown, believes that "the [tax] deduction idea could be adopted to a eugenics program by limiting deductions for couples with inferior genotypes and increasing the number of deductions for couples with superior genotypes."[98] Vukowich feels that sterilization of persons after they have produced a "legal" limit of children would be an excellent way to gain compliance with a eugenics campaign. The professor concludes with an appeal for governmental control over eugenics policies: "The selection of desirable and undesirable traits can be left to the legislature. Indeed, legislatures have enacted eugenics laws in the past. Legislative rather than parental choice would provide greater unity of effort; it would be a more effective means of diminishing detrimental genes and propagating superior ones than parental choices that could vary from couple to couple."[99]

The new eugenics movement in America, like its predecessor sixty years ago, has the enthusiastic support of some of the most distinguished scientists alive. It is no trivial matter when scientists of the stature of Nobel laureate Sir Francis Crick remark: ". . . no newborn infant should be declared human until it has passed certain tests regarding its genetic endowment and . . . if it fails these tests it forfeits the right to live."[100] Nobel laureate Linus Pauling has even suggested that there should be tattooed on the forehead of every young person a symbol showing the person's genotype. "If this were done," says Pauling, "two young people carrying the same seriously defective gene in single dose would recognize the situation at first sight and would refrain from falling in love with one another."[101] Pauling says that he favors legislation along these lines.

In looking for a statement that best sums up the new view of some scientists concerning the future and the role of eugenics and genetic engineering, the words of geneticist James F. Danielli, director of the Center for Theoretical Biology, SUNY at Buffalo, stand out. Danielli argues that "from the view of genetics, man is a barbarian, clad in the trappings of a civilization in which he is ill at ease and barely able to contend . . ." Danielli goes on to say that "most men contribute little to civilization other than maintenance services (if that), and are consumers rather than originators of civilization." According to Danielli, with human genetics as it is, the possibility of improving the human condition by improving human institutions and environments "is a dubious proposition." Danielli believes that it is now time to "consider other possibilities if civilization is to persist and advance to a modestly stable state." The other possibilities, says Danielli, "lie in genetic engineering."[102]

3
life in the laboratory

WHEN SCIENCE FICTION BECOMES SCIENCE FACT

> When I was working with H. G. Wells on the film *Things to Come* some 40 years ago, we had great difficulty designing the dresses for our future, but we never contemplated redesigning the people who would wear those dresses. Today, however, there are those who would like to do so. Do not think this is science fiction. The trouble with fantasy writers today is that they are inevitably overtaken by tomorrow's facts. The time scale of change has been recalibrated so that tomorrow is already here.[1]

On the following pages, we will embark on a unique journey through the laboratories that form the world of the bioengineers. There we will glimpse the dawn of the Genetic Age, a future reality based on the melding together of eugenics philosophy and sophisticated biological technology. Like the journeys of the past, it is marked by the intermingling of philosophy and technology, ideology and biology. This time, however, it is the technology itself, so impressive in what it can accomplish, that has made this excursion so qualitatively different from all those that have preceded it. We have now moved from promise to performance, and in so doing have traveled the first of many steps into a world where social institutions become a reflection of our inner biology and human beings the product of carefully determined genetic engineering; a world where efficiency becomes the measure of all things; where the diverse, the random, and the vagaries of chance are all set by the wayside on the road to biological progress.

The ideal of this new order of the ages is manufactured life. The architects of this future are skillful genetic technicians who collectively will assume the position of "business manager for the cosmic process of evolution."[2] Dr. Joshua Lederberg suggests we call their *modus operandi* "algeny."[3] Just as the medieval mystics attempted to turn lead into gold through the manipulations of alchemy, so in the future the genetic engineers will seek to transform leaden human inheritance into a golden biological treasure trove.

How will this technologizing of life alter biological reproduction as we know it? Professor Donald Fleming, a Harvard historian, believes the answer to that question might best be ascertained by understanding our bio-future as an extension of a contemporary industrial assembly line. In both manufacturing processes, he points out, "the number of units to be produced is a matter of rational calculation beforehand and tight control thereafter. Within certain tolerances, specifications are laid down for a satisfactory product. Quality-control is maintained by checking the output and replacing defective parts. After the product has been put to use, spare parts can normally be supplied to replace those that have worn out."[4]

Comparing traditional "organic design" (natural selection) with the prospect of laboratory-controlled life, Dr. George Wald observes, "Technological design begins with specifications and then tries to realize them. Natural selection has no specifications, it is a system of producing endless variations and then selecting out of them those that work a little better." The difference between the two, Wald contends, is that organic design is a process of "editing" while "technological design is a process of authorship." If we are embarking on the course of turning life over to technological design by selected specification, the obvious question that arises is "Who is going to set those specifications?"[5]

Some observers worry that the very technology that ushers in the Genetic Age will inevitably violate the human spirit. The renowned ethicist and author Dr. Leon Kass argues that "increasing control over the product is purchased by the increasing depersonalization of the process."[6] Jacques Ellul, the French philosopher and social critic, predicts that we will eventually become overwhelmed and consumed by the biological technology we create: "When technique enters into every area of life, including the human, it ceases to be external to man and becomes his very substance. It is no longer face to face with man, but is integrated with him, and it progressively absorbs him."[7] Pursuing Ellul's line of thought, Dr. Salvador Luria asks, "When does a 'repaired' or 'manufactured' man stop being a man . . . and become a robot, an object, an industrial product?"[8]

Proponents of human genetic engineering, on the other hand, believe questions of this sort reveal more about those who ask them than the technology in question. "To be civilized," avows Joseph Fletcher, "is to be artificial, and to object that something is artificial only condemns it in the eyes of subrational nature lovers or natural law mystics."[9] Professor Colin Austin is not as blunt but is no less forceful when he says that "if we are to be masters of our fate, we must surely accept the responsibility of exploiting to the full all advances in the control of human reproduction and development."[10]

Austin's words could be inscribed as the motto of the bioengineers in hundreds of laboratories around the world. Quietly, but unrelentingly, the artificial biological future that he and Fletcher herald is nearly upon us before we have realized its very existence. It is one of the great social ironies in our history that at a time when the President of the United States publicly urges those "living in sin" to take the vows of marriage, movie stars and publishers are convicted of peddling pornography, and relatively minor sexual affairs drive the powerful from Capitol Hill, sex, at least biologically, is being made passé by genetic engineering.

"Where do babies come from?" is a question that continues to bedevil embarrassed parents. Now, thanks to the new processes of manufacturing life, there are at least eight answers to that innocent question, not counting the one most of us are familiar with. A list of "potential technologies of eugenics and euphenics" compiled by Lederberg includes: (1) artificial insemination of a wife by her husband; (2) artificial insemination by a donor; (3) ovarian, or egg, transplant from one woman to another and subsequent artificial insemination by either the husband or a selected donor; (4) fertilization of an egg in vitro (in glass) followed by implantation into a woman; (5) extracorporeal gestation (test-tube baby); (6) parthenogenesis (virgin birth), the development of an unfertilized egg; (7) nuclear transplantation, or cloning, in which a cell is made to produce an exact genetic duplicate of the donor of the nucleus, male or female; (8) embryo fusion, or the joining together of the two individual embryos to form a human with four biological parents instead of the traditional two.[11]

Not all of these methods are now feasible, though extensive experimentation has proceeded on each. Lederberg, for one, believes that, given enough money, "essentially anything that we care to do in the area of biological engineering" can be accomplished in the next ten to thirty years.[12]

These alternatives to the old-fashioned method of sexual intercourse ("forms of rational artificial assistance to human reproduction," in Fletcher's words[13]) gain their real significance when combined with the molecular knowledge resulting from the Biological Revolution.

Artificial insemination, embryo implants, cloning, and test-tube babies, for instance, may result in full-scale genetic engineering, a government report indicates, because they open the way "to substitute the synthetic DNA for the natural DNA in the egg, sperm, or body cell and thus regulate the development of the individual."[14]

Since the easiest path to the Genetic Age is by shaping the biological inheritance of future generations, the wedding of molecular biology with the methods of manipulated birth produces a powerful marriage of eugenic interest. In this combination, suggests the Journal of the American Medical Association, lies the future:

> The popular term, genetic engineering, might be considered as covering anything having to do with the manipulation of the gametes or the fetuses, for whatever purpose, from conception other than by sexual union, to treatment of disease in utero, to the ultimate manufacture of a human being to exact specification. . . . Thus, the earliest procedure in genetic engineering . . . is artificial insemination, next . . . artificial fertilization . . . next artificial implantation . . . in the future corporeal gestation . . . and finally, what is popularly meant by genetic engineering, the production—or better the biological manufacture—of a human being to desired specifications.[15]

As with all fields of technology and experimentation, a question that confronts us at each step is: Where do we draw the line? Leon Kass has noted that in this field "one technical advance makes possible the next and in more than one respect, the first serves as a precedent for the second, the second for the third—not just technologically, but also in moral arguments. At least one good humanitarian argument can be found to justify each step."[16]

Joseph Fletcher, along with many other of the harbingers of the Genetic Age, argues that there is no clear point where we can say "no" to the new eugenics. "If the greatest good of the greatest number (i.e., the social good) were served by it," he says, virtually any reason for the modification of humanity would be justifiable, including the necessity "to bio-engineer or bio-design para-humans of 'modified men.' "[17]

Fletcher's concept of "situational ethics," a world view wherein nothing is immutable or inalienable, is in reality not "situational" at all, but simply a rephrasing of a technological imperative that underlies all of modern science: *If something can be done, then it should be done.* As we shall see on the following pages, the technological imperative is the guiding principle of genetic engineering. Each technique or process seems to begin rationally enough: a husband is sterile; he and his wife want a baby; she goes to a clinic to be artificially inseminated by the sperm of an anonymous donor. But then the inner logic of genetic engineering carries the technique forward to its final conclusion: Certain "superior" types should be selected as sperm donors; national screening programs should be used to determine whether it is "desirable" for society to allow certain individuals to reproduce; children should not be the result of sexual procreation, but should become our well-planned creations, manipulated chemically before birth to produce optimal qualities. The old philosophy of eugenics, it seems, inevitably lurks behind every genetic aid.

Genetic visionaries are already anticipating modifications that will move the human species far beyond itself on the evolutionary ladder. With such a grand scheme in mind, it is no wonder that the researchers we shall meet on the following pages have few qualms at all about the bioengineering that is already taking place in their laboratories. Their goal is simply the redesigning and control of life. And we are the objects of that control.

THE SPERM MERCHANTS

On August 18, 1971, the *New York Times* carried an article about a young man, age seventeen, health excellent, an A student. He was, by all accounts, average in every respect—except for one: This teenager was the first human being conceived from sperm that had been frozen, stored, thawed, and finally injected into a woman's uterus, where it impregnated an egg and developed to full term.[18]

The anonymous subject of this *New York Times* story is just one of perhaps a million living Americans who were born through the use of artificial insemination. The number is increasing at the rate of more than twenty thousand a year.[19] If you are surprised at these large numbers, you are not alone. Less than ten years ago, a public-opinion survey conducted by pollster Louis Harris found that only 3 percent of the American public had even heard of artificial insemination.[20] Very likely, the vast majority of those who were conceived by the method not only do not know the facts of their own birth, but are probably as ignorant of artificial insemination as most of us are.

Close to 1 percent of all children born in this country are the product of artificial insemination. The technique falls broadly into two categories: AIH (artificial insemination by husband) and AID (by donor). Couples turn to AIH when the husband has a low sperm count in his semen, or is sexually impotent but has normal sperm quantity and quality. AID is often employed when the husband is absolutely sterile, or has a hereditary disease which might be passed on to his children, or when a blood incompatibility exists between husband and wife.

AIH and AID are not limited solely to these applications, however. Because sperm is so readily available, it has always been a favorite object of manipulation by those who seek to direct human breeding toward the production of "better" individuals. From Plato's day until very recently, eugenicists have largely had to content themselves with the sterilization of those who possess "inferior" sperm while maximizing the potential of "superior" types to sow their seeds as widely as possible. The advent of artificial insemination and, more recently, the establishment of frozen-sperm banks represent a great leap forward to those who wish to breed a better race. Today, a preselected individual can have his semen stored indefinitely and used to impregnate literally millions of women, even decades after his death. Potentially, AID represents the old eugenics dream come true.

Artificial insemination is the outgrowth of three hundred years of research on human sperm. The field of investigation began in 1676, when a British medical student, using one of Anton van Leeuwenhoek's newly invented microscopes, peered through a lens at a sample of human semen. There he saw for the first time spermatazoa, minuscule creatures consisting of nothing but an oval head and a long, whiplike tail. Leeuwenhoek went on to examine semen from a number of different species, finding the tiny creatures each time. In 1685, he even managed to extract living sperm from the uterus of a dog following coitus.

At the time, however, no one knew what these squiggling entities did; it would take nearly two hundred years until the union of a sperm and a female egg was first witnessed. A few enlightened souls of the period ventured that sperm might have something to do with pregnancy. A theory they advanced held that the head of a human sperm contained a miniature human being, who would uncurl and begin to grow once inside the female. One researcher went so far as to claim that he had actually seen a tiny horse crammed into a drop of horse semen.

We've come a long way in our knowledge of the sperm since 1676. According to Dr. Don W. Fawcett of the Harvard Medical School's Laboratory of Human Reproduction and Reproductive Biology, "There's been more progress in our understanding of the spermatozoon in the past 20 years than in the first 280 years following Leeuwenhoek's discovery."[21] We now know that a human ejaculation contains approximately 300 million active sperm, each about five hundredths of an inch long. So tiny are these male sex cells that a single thimble could hold over three billion, or enough to double the population of the world. Far from containing a microscopic person, the head of the sperm is actually composed almost exclusively of twenty-three tightly packed male chromosomes. This number represents half of the chromosomal make-up of a human. When sperm are ejaculated into the uterus, they use their tails to "swim" toward their destination, an egg contained in one of the female's two oviducts. This egg also contains twenty-three chromosomes, half of the female's genetic complement. If a variety of conditions are right, one of the original three hundred million sperm will penetrate the egg, their two sets of chromosomes will merge, and the egg will begin to divide, ultimately forming a new life.

It is not altogether clear when people first began to alter this natural biological process. An Arabian manuscript dated 1322 mentions a horse breeder who stuffed a piece of wool in a mare's uterus overnight. The next day, he removed the wool, thrust it under the nose of a prize stallion, and then caught the horse's ejaculated seed on the wad. He returned to the mare and reinserted the wool, and a foal was duly born.[22]

A more scientific approach to artificial insemination was pioneered in 1776 by an Italian biologist, Lazzaro Spallanzani. As Jefferson, Adams, and Franklin were shaping the Declaration of Independence in Philadelphia, on the other side of the Atlantic Ocean Spallanzani was carefully placing miniature oilskin breeches on a male frog and then setting the pantalooned amphibian loose to try to mate with a female. Following the attempted intercourse, he undressed the frog, scraped out the milky substance he found in the trousers, and poured it over a batch of unfertilized frog eggs. The eggs went on to develop into healthy tadpoles, and the first laboratory insemination was fact. Spallanzani followed up his frog experiments by performing the same feat with neighborhood dogs.[23]

Today, artificial insemination of many kinds of domesticated animals has virtually replaced sexual intercourse. Presently, 95 percent of all cattle born in the U.S. (over 60 million cows) are produced by artificial insemination, using sperm supplied by a limited number of prize stud bulls. Some 50 million ewes, 1 million sows, 125,000 mares, 60,000 goats, and 4 million turkeys regularly give birth through AI.[24]

Though there are indications that the first human artificial insemination was accomplished about the time of the Civil War, the first verified case seems to have been in 1884. It was, without a doubt, one of the least glorious launchings of a new technique in medical history. A woman with an unspecified ailment visited her physician, a Dr. William Pancoast. Putting his patient under anesthesia, Pancoast proceeded without her knowledge to artificially inseminate her with semen obtained from one of the student interns in attendance at the hospital. Nine months later, a healthy child was born. Pancoast eventually confessed his indiscretion to the woman's husband, who apparently asked only that the doctor not reveal the matter to his wife.[25]

Since those early days, artificial insemination has moved out of the back room and into the clinical laboratories. A significant breakthrough in the field came in the 1940s, when Dr. A. S. Parkes serendipitously discovered that glycerol could be used to deep-freeze and store sperm for later use. After extensive experimentation on animals, Drs. Raymond Bunge and Jerome Sherman of the State University of Iowa became the first to utilize frozen sperm in human women. The resulting offspring were as normal as any other babies. Today, doctors like S. J. Behrman keep sperm frozen for years at a time. Noting that he has personally performed more than one hundred successful artificial inseminations of women using frozen sperm, Behrman says "there is every reason to believe that this suspension can be prolonged indefinitely."[26]

This ability to deep-freeze human sperm for virtually any length of time has fueled a budding national industry: the commercial sperm bank. Currently, banks are located in twelve cities, including Chicago, Los Angeles, New York, Omaha, and St. Paul. Times have never been better in the sperm-bank business. According to Dr. Marie Pichel Warner, a pioneer in the field, "there has been a significant increase in the use of artificial insemination. I would say there are four times as many cases" in 1976 as in 1972. Adds Dr. John Olsen, president of Cryogenic Laboratories of St. Paul, "in the last two years the number of physicians requesting and using frozen donor sperm has doubled. We're having a difficult time keeping up with the requests coming in." Olsen's firm supplies doctors in forty-five cities around the country.[27]

Cryogenic Laboratories, in fact, is a national leader in sperm storage. Cryogenic has two types of customers; those who are storing and those who want to purchase. For an initial fee of fifty-five dollars and an annual frozen-storage payment of eighteen dollars, any man can walk into the company's office and have his sperm banked. Generally a depositor will bring a sperm sample with him, though on-site facilities are available at most banks, complete with comfortable armchair and an assortment of "men's magazines." The sperm is diluted with glycerol, deposited in a dozen or so tiny bottles, which are then placed in metal containers. These are submerged in liquid nitrogen and frozen to a temperature of –321°F. Olsen says that many of his clients are men who store their semen prior to vasectomy. Cryogenic also handles "a lot of professional athletes and high risk professionals and people undergoing radiation therapy."[28] Occasionally a client has a more exotic motivation. One prominent Minnesotan has frozen his sperm, it is reported, so that should anything happen to his only son, his daughter-in-law could preserve the family line by using his semen.[29]

As with banks that deal with money storage, the depositor at Cryogenic can withdraw his sperm at any time through his doctor. On some depositors' accounts, a wife also has withdrawal privileges. Recently some doctors have reported that they are artificially impregnating wives whose husbands must be away from home for long periods of time. There have also been reports that wives have used their husband's sperm to create a child after the spouse's death. Undoubtedly, in the future the ability of frozen sperm to bridge the gap of separation of man and woman by time and distance will be put to increased use.

Cryogenic spokesmen are careful to indicate that in the case of a man who deposits his sperm for his own personal use, "our policy is that we do not own the semen: we are just a humble storage facility." Dr. Robert Ersek, medical director of the firm, puts it this way: "We don't want to be in the position of owning semen and controlling who gets it. [If that happened] we could get into the difficult position of, say, owning

Einstein semen and auctioning it off to the highest bidder."[30]

In a sense, however, Cryogenic and other commercial firms do something very similar when it comes to their other category of client: women who wish to purchase sperm for AID. The sperm used in AID by Cryogenic is obtained from just thirty-five donors, all "well scrutinized." Olsen claims that "not anyone can come in off the street to be a donor. Seventy-five to eighty percent of donor applications are refused during a screening process which includes three generations of testing, blood tests, etc." As part of the screening donors must fill out forms listing physical characteristics, as well as information regarding occupation, religion, and IQ.[31]

When Cryogenic is contacted by a doctor on a woman's behalf, the company gives the prospective client a "donor search form," which lists various characteristics of the company's sperm donors. The woman ranks the characteristics in order of importance, and the firm attempts to match her specifications to its stored sperm. A firm sperm-bank policy is that the recipient of the sperm can never meet the donor.

Recently, a new trend in AID has begun to take hold. For the first time, single women are approaching sperm banks requesting artificial insemination. Dr. Wayne Decker, Surgeon in Chief of the New York Fertility Research Foundation, says the number is small, "less than 20 AID's a year for single women" at his clinic, but nonetheless significant.[32]

As for socioeconomic background of patients who seek artificial insemination, most are college-educated or professionals. Dr. Wilfred J. Finegold, head of the Division of Sterility, Planned Parenthood Center, Pittsburgh, and author of *Artificial Insemination,* believes the reason for this is that "It is an expensive procedure, since 2 to 3 inseminations a month are necessary, and these may be required for as long as six months before a pregnancy is achieved."[33]

As a technique of eugenics, AID is simply an extension of well-established and widely used animal-breeding programs. The potential mother attempts to find a donor who embodies certain physical traits that she would like to see reproduced in her child. At firms like Cryogenic, a client can even go beyond biological specifications, by giving her baby the supposed benefits of a father with the proper occupation, religion, and intelligence level. Observers like bioethicist Marc Lappé believe that this process, though strictly voluntary, "is functionally serving as fostering eugenic aims" by initiating an unhealthy preoccupation with the "quality" of AID sperm.[34] A government study entitled *Genetic Engineering: Evolution of a Technological Issue* noted in 1972 that conceptually AID might be laying the foundation for widespread "eutelegenesis," or artificial insemination by semen from a donor carefully selected for specific characteristics.[35] Eutelegenesis is, in fact, the motivating force for many advocates of the wider use of sperm banks. During a forty-year period, the late H. J. Muller, a man often credited as the "father of frozen sperm," proselytized in favor of the establishment of national sperm banks, which all parents—not just those suffering from infertility—would be encouraged to use. This program of "germinal choice," he said, would make widely available the sperm of those blessed with a "depth and scope of intelligence, curiosity, genuineness and warmth of fellow feeling, the feeling of oneness with others, joy in life and in achievement, keenness of appreciation, facility in expression and creativity."[36]

Whether germinal choice will catch on in the future is open to question. If it does, the process for selecting those individuals with the proper talents and characteristics should prove fascinating. Richard Nixon, for instance, might be eligible because he has a certain "facility in expression," but his "feeling of oneness with others" might not be up to standard. Most other individuals would probably come up short, too. After all, as the genetic engineers constantly remind us, no one is perfect. Because of this, eutelegenesis will undoubtedly be combined with more modern biotechnologies. Since the head of a sperm is little more than a bundle of chromosomes, gene manipulation might be employed to modify the genetic information before artificial insemination. The resulting child could then be truly manufactured, complete with specification and quality control.

There are more immediate eugenic applications of artificial insemination, however. Many experts believe that by the end of this decade parents using AI will be able to choose their baby's sex with a 90 percent degree of accuracy. Each sperm carries within its chromosomes genes that indicate sex. The Y chromosome of the androsperm will make a male; an X chromosome, or gynosperm, a female. Tests have shown that male sperm is smaller, lighter, and speedier than the female. Using fluorescent-lighting techniques, the two types can be differentiated in the laboratory. One clinic which has managed to separate androsperm and gynosperm is already accepting applications from couples who want male children only.[37]

Do parents want to predetermine the sex of their children? Apparently they do. One study, conducted in 1973 at the Stanford Medical Center, indicated that 36 percent of parents would select the sex of their child if given the opportunity, while another 31 percent said they might do so but were not sure.[38] This possibility may present unexpected social problems. In a society with as much sexual inequality as ours, asks science writer Caryl Rivers in *Ms.* magazine, "are we ready to assume the responsibility of sex selection?"[39] Rivers and others point to surveys that have consistently shown that a substantial majority of parents would choose a male offspring, at least for their first child. Thus, sex selection through artificial insemination may dramatically upset the natural process which keeps our sexual division, if not completely equal in numbers, at least on an even keel.

The continued spread of AID may well pose other social problems. Commercial firms like Cryogenic use a few dozen sperm donors to inseminate thousands of women. Might this result in unwitting incest between different AID children who have no idea they share the same biological father? As fanciful as this may seem, *Time* magazine reported several years ago that a New York physician had to prevent the marriage of two young people for this very reason.[40] If AID keeps up its phenomenal growth, couples in the year 2000 may be consulting sperm data banks and semen registries before taking their marriage vows.

What effect would widespread AID have on family relationships? What new forms of sibling rivalries will develop among children who share the same mother but have different semen donors for fathers? How will children relate to the male figure who is present in the home but has no biological connection with them? Will children blame their parents for making a poor selection of sperm? Will families suffer from an unstated obsession to learn the identity of the semen donor?

As a forewarning of the monumental social problems posed by the advance of the Genetic Age, our judicial system has not yet fully addressed the many legal questions arising from AID. The legitimacy of children born of artificial insemination remains in doubt, even though one in every two hundred Americans falls into this newest of minority groups. While courts in the states of Georgia, Kansas, Oklahoma, New York, Arkansas, California, Maryland, and North Carolina have ruled such children legitimate, others have declared just the opposite. In many states AID is still considered to be adultery and the woman who is inseminated an adulteress. Who, then, is her partner in adultery? The anonymous donor of sperm, whom she will never meet? The doctor who injects the semen into a patient's uterus? If so, can a female doctor commit adultery with her female client?

One widely cited case that involved many of these very elements is that of Mr. and Mrs. John M. Prutting. Mr. Prutting had been "absolutely sterile" for some eighteen years as a result of exposure to radiation in the course of his work. Without her husband's knowledge or consent, Mrs. Prutting was artificially inseminated by her physician. After she gave birth, her husband sued for divorce on the grounds of adultery, claiming that she had voluntarily surrendered herself to another person.[41]

Another legal battle raged over a child who was conceived by AID with the husband's permission. Several years later, the couple separated but the husband wanted visitation rights to see the child. The wife argued that her husband had no such right because he wasn't the child's biological parent. He countered by noting that he had given his permission to the birth and had helped raise the child as his own. A New York court granted the man visiting rights, but the woman and her child moved to Oklahoma, where a court reversed the New York finding on the grounds that the husband was not the child's genetic father.[42] This ruling raised more questions than it answered. If the woman's husband was not the child's father, then who was? The donor of the semen? If so, could that donor arrive in town one day and demand to visit his child? If the woman happened to learn the donor's identity, could she sue him for child support? Could the child one day claim to be the donor's heir? Strange questions, to be sure—but very real for one million Americans.

EGGS, EMBRYOS, AND TEST TUBES

While one group of researchers concentrates on manipulation of the sperm, a second and related field of investigation centers on the female sex cell, the ova or egg. The technologies surrounding the egg—ranging from embryo transplants and the storage of frozen embryos to in vitro (in glass) fertilization and test-tube life—represent a far greater degree of eugenic sophistication than that associated with artificial insemination.

All forms of egg manipulation require a handy supply of ova for experimental purposes. Today eggs, both fertilized and unfertilized, are routinely obtained from a wide range of mammals, including humans, through the use of "super-ovulation" hormones. The process of removing already impregnated eggs from one animal and implanting them in the uterus of another from the same species is already accomplished regularly. Dr. M. C. Chang, one of the leaders in embryo "flushing," has conducted an extensive study of the techniques using mice, hamsters, rabbits, and rats.

Recently, animal breeders have combined artificial insemination with embryo flushing and implantation to create super-herds of prize cattle. Normally a cow will produce just one egg at a time for fertilization, though she has the physical capability of producing many more. Now, using selected hormone injections, breeders can trick the cow's reproductive mechanism into releasing as many as thirty eggs at a time. Once this has been accomplished, the breeder artificially inseminates his cow with semen taken from a stud bull. The result is a "super-embryo," the product of the eggs from a prize cow and the sperm of a prize bull. After a few days of gestation, the embryos are flushed from the cow and implanted in thirty other cows, each quite healthy but from inferior stock. The calves that are born are genetically the offspring of the original donors of the egg and sperm and bear no resemblance at all to the cows that carried them.

While this process might seem bizarre to those outside the cattle industry, breeders are reaping hefty profits from embryo transplants. According to Casey Ringleberg of Modern Ova Trends, a commercial breeding firm, each calf produced in this manner is worth up to three thousand dollars. Animals from superior strains like Limousin, Simmental, and Chianini may sell for as much as one hundred thousand dollars. Many cattle-embryo companies use sophisticated computerized registries to guarantee the quality of a product that is literally worth its weight in gold.[43]

Dr. E. S. E. Hafez, chairman of the Department of Animal Sciences at Washington State University, believes that someday transplant techniques will be used to ship entire herds of super-embryos around the globe. To prove his point, Hafez, whose work is generously funded by the National Institutes of Health, had a friend in England send him one hundred prize sheep in a small box carried on a commercial airplane. The sheep, in the form of newly conceived embryos, made the journey as passengers implanted in the womb of a rabbit, where they lived and developed for several days. When he received the package, Hafez simply opened the rabbit, removed the embryos, and had them implanted into normal American ewes. Months later, the doctor was the proud owner of a prize herd of English sheep.[44]

More sophisticated is the use of deep-frozen embryo banks that allow for long-term storage of super-embryos. The first success in this area occurred in 1972 through the work of Drs. Peter Mazur and David G. Whittingham of the Atomic Energy Commission's Oak Ridge Laboratories in Tennessee. Late that year the two scientists announced that frozen mouse embryos had been successfully thawed and implanted. Their experiment involved the freezing of 2,500 mouse embryos to the temperature of –452°F. After eight days in the freezer, the embryos were slowly thawed. Of the 1,000 that survived the warming process, more than 400 came to full term. To prove that the implanted embryos were indeed the resulting offspring, Mazur and Whittingham placed thawed but otherwise normal embryos in albino mice. All of the babies had dark coats and eyes.[45]

Since that time, embryos flushed from mice have been frozen for more than a year with a survival rate of up to 80 percent. In announcing their scientific first, the two researchers suggested that their work might lead to the freezing of human organs for later use, along with embryos of animals "of stock with an optimal genetic background." Others have suggested deep-freezing embryos of endangered species, preserving them until the day when a functioning artificial womb is developed. (The AEC, it should be noted, was careful to issue a statement that it did not expect its experiments on mouse embryos to be extended to deep-frozen human babies.[46])

At about the time of the AEC's technical triumph, the Unit of Reproductive Physiology and Biochemistry of England's Agricultural Research Council proudly announced the birth of Frosty, a bull calf born from a deep-frozen embryo. As with the frozen mice embryos, Frosty had been removed from the uterus of his biological mother, frozen in liquid nitrogen, thawed, and implanted in another cow. Cattle breeders expect this development will revolutionize their industry.[47]

It may also revolutionize human reproduction. At least, that's what major institutional financial backers of such research—the National Institutes of Health, the Population Council, the Ford Foundation—along with many of the researchers, expect. Scientists like Dr.

Ralph Brinster, of the University of Pennsylvania's School of Veterinary Medicine, point out that there appear to be considerable similarities in embryo implantation and development between mammals like mice or rabbits and other species, including humans.[48] In fact, the new methods of selective animal breeding are already being extended to humans. The research has mushroomed into a heated competition to achieve major "firsts" in the manipulation of human reproduction. One doctor has called it "the embryo sweepstakes."[49]

Commercial human-ova banks, for instance, are right around the corner. Several egg depositories already exist at universities. Dr. James L. Burks of the University of Chicago has bridged the gap between his work on animals and humans: He has deep-frozen both rabbit eggs and human ova. When thawed, these eggs appear to be viable and suitable for implantation. Though Burks cautions, "we are not trying to grow people in test tubes," he does envision transplanting thawed ova from one woman into another who is infertile.[50] In this way, the second woman could have a baby, though it would have no genetic link to her. The possibility of egg transplants, like AID, raises a host of provocative questions, not the least of which is, who would be the mother of the newborn baby, the woman who contributed the egg or the woman who gestated it?

So far a human-egg transplant has not been attempted, though the technical capability is at hand. Several years ago, a prominent British gynecologist was set to conduct such an operation, but he backed off when an English medical organization informed him that under current statutes the resulting child would be illegitimate.[51]

A related area of experimentation is the fertilization of an egg in the laboratory and its subsequent implantation into a woman, a technique which James D. Watson feels is "likely to be general medical practice, capable of routine performance in many nations, within some ten to twenty years."[52] Ultimately this line of research should lead to the creation of test-tube-grown babies.

The ability to keep tissue alive in vitro has been recognized since 1907, when Dr. Ross Harrison of Johns Hopkins University began growing frog cells in petri dishes. In 1912, Dr. Alexis Carrel, a Nobel Prize–winner, removed a piece of tissue from a chicken heart and placed it in a glass container filled with fluid. Carrel's chicken heart continued living in vitro for a full 33 years, outliving the scientist by two years. Carrel eventually came to regard this feat with such awe that he began insisting that his lab technicians wear flowing black robes and hoods whenever in the presence of his biological icon. More recently, a woman named Helen Lane achieved a kind of immortality, though she has been dead for a number of years. While treating her for cancer, Lane's doctor removed some diseased tissue for analysis. Though cancerous, her cells proved extremely hardy, and samples of them, called HeLa cells, continue to be in use in laboratories around the world.[53]

The fertilization of eggs in a laboratory is a development related to the tissue-culture field. As far back as the 1940s, Dr. John Rock, in the course of experiments to develop a birth-control pill, succeeded in fertilizing a human egg with sperm in a test tube. Following fertilization, the egg went through several divisions. A decade later, Dr. Landrum Shettles of Columbia Presbyterian Hospital repeated Rock's success. His microphotographs of the union of an egg and a sperm in vitro appeared in dozens of medical textbooks. But, like Rock's, his embryo lived only a short time.[54]

In 1961, an Italian biologist, Dr. Daniele Petrucci, announced he had had considerably greater results than his predecessors. In a still-controversial experiment, Petrucci fertilized an egg and allowed it to grow outside the human body for fifty-nine days. He claimed that "a heartbeat was discernible" but he was forced to destroy his creation because "it became deformed and enlarged—a monstrosity." Petrucci, a Catholic, went so far as to give his embryo conditional baptism and extreme unction, but this was deemed insufficient by Italian authorities and the Church hierarchy. Under orders from the Pope, Petrucci finally gave up his attempts to grow life in vitro.[55]

Petrucci's baby created a worldwide sensation that even reached into the Soviet Union and the People's Republic of China. *Jenmin Jin Pao,* the official Chinese newspaper, praised his effort: "These are achievements of extreme importance, which have opened up bright perspectives for similar research. . . . Nine months of pregnancy is no light or easy burden and such diseases as poisoning due to pregnancy are detrimental to health. If children can be had without being born, working mothers need not be affected by childbirth. This is happy news for women."[56]

Soviet scientists invited Petrucci to come teach his methods in Russia. During his two-month stay at the Soviet Institute of Experimental Biology, the Italian received a Soviet medal for his work and presumably gave his hosts the benefit of his experience. By 1966, Dr. Pyotr Anakhin of the Academy of Medical Sciences in Moscow announced that he and a team of Soviet doctors had surpassed Petrucci's record by keeping alive more than 250 human embryos in test tubes for periods of up to six months. The largest specimen reached the weight of one pound two ounces, before dying.[57]

Two British doctors, R. G. Edwards of the University of Cambridge and Patrick Steptoe of Oldham General Hospital, are currently regarded as the leaders in the field. Having mastered laboratory insemination, the two have now progressed to the next logical step: implanting a fertilized egg into a uterus.

To obtain the eggs for fertilization and implantation, Edwards performs a minor operation on a patient.

Making two small slits in a woman's navel, he lowers into her peritoneal cavity a narrow tube, called a laparoscope, a device that serves as a combination searchlight/telescope. Once he has located her ovaries, Edwards inserts a second narrow tube, which sucks out eggs from the follicles on the ovarian surface. A few stitches close the hole, and after a single night in the hospital the patient is sent home.[58]

Edwards then initiates fertilization using the sperm of the husband of the egg donor. A few days later, his patient returns to the clinic. A tube is inserted into her uterus and the embryo is implanted in the uterine wall without surgery. Edwards and Steptoe have attempted this procedure on a number of occasions. So far, they have encountered complications involving hormonal imbalances that abort the fertilized eggs. Their longest success has been a thirteen-week pregnancy.[59]

Others have also entered the "embryo sweepstakes." In 1973, Dr. Landrum Shettles, assisted by Dr. William J. Sweeney of the Cornell Medical School, removed an egg from an infertile woman, Doris Del Zio. Fertilizing the egg in vitro with the sperm of Mr. Del Zio, Shettles prepared for what he hoped would be the first successful human embryo implant. Before the operation, however, his superior at the hospital, Dr. Raymond Vande Wiele, removed the developing embryo from its glass container and destroyed it. Vande Wiele claimed that further experimentation on lower primates was needed before the procedure could be safely attempted in humans. Shettles resigned in protest, later establishing the nonprofit New York Fertility Research Foundation to continue his work. The Del Zios sued Dr. Vande Wiele for destroying their potential baby.[60]

About the time that Shettles was preparing to perform his operation on Mrs. Del Zio, James Watson appeared before a Senate committee in Washington to warn that the first baby resulting from an embryo implant could be born at any time. When it happens, he added, "all hell is going to break loose." In fact, all hell did break loose in 1974 during a British medical conference when one of the featured speakers, Dr. Douglas Bevis, stunned his audience by glibly stating that after thirty attempted implantations, three had finally been "crowned with success." His announcement caused a storm of controversy that rages to this day. Because of the carnival atmosphere generated by the world press coverage of his work, Bevis has refused to discuss his experiments except to say that two of the babies are currently living in England and the third in Italy. Though one newspaper offered him seventy-two thousand dollars to tell all, Bevis remains silent. One day, he says, he will explain his work in detail in the pages of a medical journal.[61]

As the day of widespread embryo transplants nears, public concern for the meaning of the work increases. Edwards is the first to admit that his research "provokes various thoughts and opinions: the beginning of test tube babies, armies of carefully planned robots [and] playing God in the laboratory."[62] So far as he is concerned, though, his motivations couldn't be further from these frightening possibilities. He and his colleagues, he says, are simply interested in helping women who can't have babies naturally because of blocked Fallopian tubes or other ailments. Embryo implants, he believes, are the only option for his patients. Dr. Shettles puts it more graphically: "If the bridge is out, what's wrong with using a helicopter?"[63]

According to Edwards, at least twenty thousand women in England could benefit from his techniques, and possibly as many as one million in the U.S.[64] "We are all aware that this work presents challenges to a number of established social and ethical concepts," he concedes, but "the emphasis should be on rewards that may be promised in fundamental medicine."[65] As a medical technique, embryo implants are impossible to halt, he argues, because they are so easy to do. To legislate against them, as some have proposed, would lead to "smuggling of the work or emigration of the people who do it."[66]

Edwards's scenario of a flourishing back-alley trade in embryo implants is certainly distasteful, but so may be some of the side effects of the work he is engaged in. An embryo is a delicate thing, doctors point out. How can we be sure that, in implanting it, we do not create a horribly disfigured baby nine months later?

Some scientists, like Dr. James Watson, have pointed out that the public will not look kindly on a researcher, and perhaps all of science, if the first birth from an embryo implant turns out to be a monstrous mistake. Watson has voiced this concern publicly, calling on Edwards and Steptoe to halt their experiments before they have gone too far. Another Nobelist, Dr. Max Perutz, derides embryo implants as a "stunt" and argues that "the whole nation should decide whether or not these experiments continue."[67]

Assuming that embryo implants can be performed safely, the technique still holds implications far beyond the treatment of infertile women. As a tool for uniting carefully selected sperm and eggs to produce some predetermined quality—"dial-a-baby," in the words of one observer—embryo implants and in vitro fertilization are a step closer to genetic engineering. The way is also being opened to the modification of life at the molecular level through the use of in vitro techniques. Noting the recent developments in the synthesis of genes, gene mapping, and the use of recombinant DNA as a tool for introducing new genetic information into cells, Dr. Bentley Glass claims these tools of manipulation "will be unquestionably most effective if carried out during the early embryonic stages of development, or in the just-fertilized egg itself before it has begun its cleavage into numerous cells. Hence we must look with expectant attention at the startling progress that is being made in the laboratory of R. G. Edwards . . ."[68]

Leon Kass speaks for many when he says "Surely, there is more at issue than providing a child for an infertile woman. Once introduced for that purpose, laboratory fertilization can be used for any purpose. Indeed, the work described is a giant step toward the full laboratory control of human reproduction."[69]

Another giant step along that road is being taken right now as biologists strive to move beyond the womb and embryo implants into the realm of life grown completely in vitro—the true test-tube baby.

BABIES UNDER GLASS

"The womb," says Joseph Fletcher, "is a dark and dangerous place, a hazardous environment. We should want our potential children to be where they can be watched and protected as much as possible."[70]

Not so very long ago, the womb was a private space. Now things are different. Today the developing fetus can be observed through the use of microphones, electromicroscopes, infrared thermography, X-rays, ultrasonics, radio-opaque dyes, and even by insertion of a miniature camera that will send out motion pictures of the growing life form to an eager public.

Soon even these prenatal-snooping techniques may become outmoded as fetologists approach "the closing of the gap from ovulation to birth in a totally synthetic environment."[71]

The late French Nobel Prize–winning biologist Jean Rostand often said that he believed the creation of the artificial womb to be "inevitable." In a grand understatement, he acknowledged that such an advent would result "in a more or less profound modification of the human being in course of formation." Until that glorious day arrived, Rostand suggested "an intermediary solution of the problem." "Delivery," he forecast, "could be stimulated artificially and the embryo placed in culture at the age of two or three months: in short, a woman could reproduce like a kangaroo."[72]

Baby-in-a-pouch notwithstanding, most researchers in the test-tube baby field are concentrating their efforts on the development of a complete artificial womb. Dr. Yu-Chih Hsu of Johns Hopkins University School of Hygiene and Public Health has succeeded in growing a mouse embryo in vitro through approximately half of its gestation period.[73] The test-tube experiments of Petrucci, Edwards, Shettles, and others have shown the potential for human fetal development in vitro.

Dr. Hafez, the biologist who specializes in flushing super-embryos from pregnant sheep and cattle and shipping them halfway around the world, is certain that the day of complete test-tube life is nearer than most of us suspect. To drive his point home, the mediagenic doctor once posed for a picture that appeared as a two-page photo spread in a national news magazine. There, in full color, was Hafez, test tubes labeled MAN, SHEEP, and SWINE held in his outstretched hand: a graphic representation of "the barnyard of the future—complete with farmer." Hafez predicts that fertilized eggs will one day be routinely deep-frozen and stored for eventual in vitro hatching, a possibility that might prove particularly useful in our national space program. "When you consider how much it costs in fuel to lift every pound off the launch pad," he says, "why send full-grown men and women aboard spaceships? Instead, why not ship tiny embryos, in the care of a competent biologist, who could grow them into people, cows, pigs, chickens, horses—anything we wanted—after they got there? After all, we miniaturized other spacecraft components. Why not the passengers?"[74]

Before this Space Age Noah's Ark blasts off, an artificial womb must first be built to house and care for a fetus during its nine months of development. Incubators can already keep alive a baby born as many as three months prematurely. The key now is to gain control of the first twenty-four weeks of the growth period, an achievement that Dr. Bernard Nathanson believes will take place within five years. Already, he points out, a few fetuses have been kept alive outside the womb for as long as eighteen or nineteen weeks, fully half of the gestation period.[75]

Dr. Robert Goodlin of Stanford has been a leader in the search for an artificial womb for over a decade. In his laboratory he has constructed a pressurized steel-and-glass world into which a very young fetus can be placed. Inside this synthetic womb, an oxygen-rich saline solution bathes the fetus. Intense pressure—roughly the equivalent felt underwater at a depth of 450 feet—literally drives oxygen through the skin of its inhabitant so that the baby's lungs do not have to work. No fetus has yet survived in the chamber for more than forty-eight hours. Goodlin is working to solve this problem by engineering a system to draw off carbon dioxide and waste materials that become deadly unless removed.

Across the country in Bethesda, Maryland, Drs. Warren Zapol and Theodore Kolobow of the National Heart Institute work at removing a lamb fetus from a ewe. Once removed, the fetus's umbilical cord is then hooked up to a device that contains a lung, pump, and nutrient supply. Zapol and Kolobow hope that such a mechanism will solve the waste-removal predicament.[76] Using a machine similar in concept, Dr. Geoffrey Chamberlain of Great Britain has sustained life in several young fetuses weighing as little as three quarters of a pound. One lived for five hours and eighteen minutes before an accident killed it.[77]

Development of the fetus completely outside the body will open the door for numerous biological modifications during the gestation period. At the Regional Primate Research Center in Oregon, doctors have already practiced fetal engineering on monkeys. The concept is simple. Doctors anesthetize a pregnant rhesus monkey, thus automatically anesthetizing the fetus within. Cutting the adult open, the surgeons remove the entire uterus without losing the precious amniotic fluid, and then lift out the unborn monkey. After electrical and biological tests are performed, the fetus is placed back in the uterus, which is finally sewed back inside the adult. The doctors who have practiced these operations are not veterinarians, but a team of surgeons from the Columbia-Presbyterian Hospital in New York. They hope to use the knowledge they gain from the rhesus monkey to conduct prenatal surgery on human fetuses. They reason that if a baby could be raised in a glass womb, necessary operations would undoubtedly be far easier to perform.[78]

Dr. Kermit E. Krantz of the University of Kansas Medical Center believes such experiments may lead to surgery that could have a dramatic impact on the future child. "It is not hard to see how any malfunction of the placenta could turn a potential Einstein into a mediocrity," he observes. "If we learn enough—who knows—maybe we can turn mediocrities into Einsteins."[79] More recently, a federal study contends that "when the techniques for fertilization in the laboratory and genetic intervention can be combined with an effective artificial placenta, the potential for genetic engineering will have been achieved."[80]

Once perfected, will in vitro fertilization and ectogenesis replace pregnancy, as Aldous Huxley predicted in *Brave New World*? Some think it a sure thing:

> Natural pregnancy may become an anachronism. The two tiny laparoscopy scars, exposed by a bikini on the beach, will be as ordinary as our smallpox vaccination, but women will no longer have lost their figures in childbearing. The uterus will become appendixlike, though the ovaries

> will be as crucial as before. At the age of 20, each girl will be able to choose to be superovulated and her eggs collected and frozen, as it is known that babies conceived by young women are less likely to suffer from mongolism and other birth defects.[81]

Joseph Fletcher, for one, sees test-tube life as finishing the work begun by the Pill. Because laboratory-created life is "willed, chosen, purposed, and controlled" rather than emotionally or accidentally entered into, Fletcher believes that "laboratory reproduction is radically human compared to conception by ordinary heterosexual intercourse."[82]

Fletcher may be right, though many parents might argue with him. What is crystal clear, however, is that test-tube life raises questions for which our current value system and thought patterns provide no ready answers. When a baby is created from anonymously donated sperm and egg in a laboratory, who are its father and mother? Does it even have any at all? Is the baby even a human being, or just the product of an imaginative experiment? Does the new life form have the rights of citizenship? What are the psychological implications of growing up as a specimen, sheltered not by a warm womb but by steel and glass, belonging to no one but the lab technician who joined together sperm and egg? In a world already populated with people with identity crises, what is the personal identity of a test-tube baby?

Some researchers say such questions are for future generations, not for ours. The technical problems of sustaining life in a test tube remain formidable, to be sure, but no more imposing than the problem of landing men on the moon appeared to be in 1961, when that became a national goal. Dr. Bentley Glass estimates that by the end if this century, just two decades away, the first test-tube baby will be "decanted."[83] Many think him far too conservative, given the incredible progress of recent years. At any rate, it is indisputably true that test-tube life will be here much sooner than Aldous Huxley envisioned. Whether it will come to have the social impact predicted by Huxley remains to be seen.

> In the Bottling Room all was harmonious bustle and ordered activity . . . The procession advanced; one by one the eggs were transferred from their test-tubes to the larger containers . . . Heredity, date of fertilization . . . details were transferred from test-tube to bottle. No longer anonymous, but named, identified, the procession marched slowly on; on through an opening in the wall, slowly on into the Social Predestination Room.
>
> The faint hum and rattle of machinery still stirred the crimson air in the Embryo Store. Majestically and forever the conveyors crept forward with their load of future men and women.[84]

WHO SHALL BE CLONED?

> . . . Given the widespread development of the safe clinical procedures for handling human eggs, cloning experiments would not be prohibitively expensive. They need not be restricted to the superpowers. All smaller countries now possess the resources required for eventual success . . . if the matter proceeds in its current nondirected fashion, a human being born of clonal reproduction most likely will appear on the earth within the next twenty to fifty years, and even sooner, if some nation should actively promote the venture.[85]

No bioengineering possibility provokes more "sci-fi" thinking than cloning. But cloning, the production of genetically duplicate individuals from the biological information contained in a single body cell, is clearly not science fiction. Within our lifetimes, carbon-copy human beings will walk among us.

This possibility, so bizarre to many of us, is in fact a realization of the long-held eugenics vision of preserving the best biological specimens. Dr. Joshua Lederberg, a leading proponent of human cloning, believes that "clonality . . . answers the technical specifications of the eugenicists in a way that Mendelian (sexual) breeding does not." Lederberg has even developed a programmatic approach to cloning that would seem the height of sensibility to the old school of eugenicists:

> If a superior individual—and presumably, then, genotype—is identified, why not copy it directly, rather than suffer all the risks, including those of sex determination, involved in the disruptions of recombination (sexual procreation). Leave sexual reproduction for experimental purposes; when a suitable type is ascertained take care to maintain it by clonal propagation.[86]

What Lederberg and others envision is the systematic extension to human beings of a phenomenon that already occurs among some plants and animals naturally. When it does, this type of cloning is called "parthenogenesis," after the Greek *parthenos,* "virgin." Natural cloning results from the development and growth of an unfertilized cell into a living organism. Because all of the chromosomes that make up this new individual are inherited from just one parent, the end product is a genetic duplicate of the original.

Numerous plants reproduce themselves in this way. Some insects, like the honeybee, the greenfly, and the wasp, give birth asexually. An armadillo will often have quadruplets that are its identical copies. Turkeys occasionally reproduce through parthenogenesis, a fact backed up by the experience of Dr. Marlow W. Olsen of the Department of Agriculture Research Center at Beltsville, Maryland. Olsen has attended the hatching of several poults which never had a father. The first such bird discovered by Olsen, named Olie, was said to prefer a hearty diet of eggs and cottage cheese. The government furnished no dietary reports on Popeye, the second turkey hatched from an unfertilized egg.[87]

Parthenogenesis has even been reported in human beings, always women. In a grown human being, every body cell, no matter where it is located or what its function, contains a full complement of the entire DNA of the body. Through evolution, our cells have learned to specialize, or differentiate, to assume various biological tasks. If a cell becomes part of a finger tip, all the other DNA information it contains is "masked." Apparently a sudden shock or bump on the head will, on rare occasion, throw on the main switch in a cell, causing all of the genetic information to manifest itself. When this happens in a woman, the cell begins to divide and develop, and may result in the birth of a child, a girl exactly like her mother in every genetic respect. Dr. Helen Spurway, a geneticist at the London University College, suggests that human parthenogenesis may cause one in every 1.6 million pregnancies.[88] Several instances of the phenomenon have been reliably documented.

One such case occurred in 1944 in war-torn Hanover. During an Allied bombing raid of the city, a young German girl collapsed in the streets. Nine months later, she gave birth to a daughter, who appeared—through blood tests, fingerprints, and other indicators—to be the exact twin of the mother. The woman swore she had not had intercourse, and extensive medical tests seemed to support her claim. Her examining physicians believe the shock of the bombing might somehow have jarred a dormant body cell within the uterus to begin reproduction.[89]

While some plants and insects regularly reproduce themselves asexually, parthenogenesis is generally a freak of nature. Cloning is not. It is directed and initiated by scientists who research some of the most bizarre terrain of the world of manufactured life. The term is derived from the Greek root meaning "cutting." In essence, this is what is done by clonal engineers. A single cell is cut, or sliced, from an organism. When properly stimulated, it can grow into a mirror image of its donor.

At the turn of the century, Dr. Gottlieb Haberlandt, an Austrian botanist, predicted that it would one day be possible to grow entire plants from a single cell.[90] In the 1950s, Dr. F. C. Steward of Cornell University did just that, using a popular American vegetable as a patient. The Steward clonal technique is begun by punching out tiny plugs from a carrot slice and placing them in a nutrient-bath solution of coconut milk and other liquids. The glass flasks that contain the solution are slowly revolved on a machine. This causes individual differentiated cells to break off. The nutrient bath then takes over. After his initial success, an astonished Steward said, "It was as if the coconut milk had acted like a clutch, putting the cell's idling engine of growth into gear." At the time, Steward admitted that he and his colleagues "were hardly prepared for such dramatic results."[91] Today Steward has adjusted. His laboratory now houses row upon row of cloned carrots, as well as tobacco plant and asparagus clones. Cloned vegetables won't be found in the grocery store for a while, though. They're considerably more expensive to grow than the real thing.[92]

The cloning of animals even antedates that of vegetables. The first reports of induced cloning of a lower animal came during the 1890s, from the French biologist Jacques Loeb. Simply by touching an unfertilized sea urchin egg with dry ice, Loeb found that he could make the egg begin reproducing, forming a new sea urchin. By 1952, one researcher reported, some 371 different methods had been found to produce sea urchins asexually through various degrees of electrical shock, chemical reaction, and biological manipulation.[93]

Readers of *Life* magazine were undoubtedly surprised in 1939 to see the cover of an issue devoted to a quite-normal-looking rabbit that had been born through deliberately induced parthenogenesis. To Dr. Gregory Pincus goes the honor of producing this rabbit through the administration of thermal-shock treatment. Pincus, who had stumbled on the process accidentally, abandoned further research into the phenomenon because it had little relevance to his major line of inquiry. He went on to do pioneering work in the development of the birth-control pill.[94]

The first cloning undertaken with sophisticated technology was performed on a leopard frog. In 1952, Drs. Robert Briggs and Thomas J. King of the Institute for Cancer Research in Philadelphia replaced the nucleus of an already fertilized frog egg with the tissue of another frog of the same species. The resulting tadpoles were all duplicates of the tissue-cell donor.

Building on the work of Briggs and King, Dr. J. B. Gurdon of Oxford has progressed a step further by using an unfertilized egg cell taken from an African clawed frog. Gurdon uses ultraviolet radiation to destroy the egg nucleus, then implants into the empty egg cells taken from the frog's intestinal wall. The egg cell, now equipped with the full chromosomal make-up necessary for it to become an African clawed frog, is evidently tricked into "thinking" it has been fertilized. The cell begins to divide. A clone is born.[95]

Researchers have yet to clone a mammal in this way. Using a less-sophisticated method, however, a just-fertilized mouse cell that has barely begun to divide can be carefully pried in two to create identical twins. Experts agree that true cloning is on the horizon. James Watson predicts that animal cloning will be particularly attractive to owners of prize cattle and other livestock: ". . . we must expect that, unless somehow strongly discouraged, veterinarians throughout the world someday will attempt the cloning of uniquely domestic animals."[96] Racehorse owners might be keen on the idea of duplicating their thoroughbreds for the annual Kentucky Derby.

Dr. John R. Platt, a biophysicist at the University of Michigan, sees no reason for us to content ourselves with the cloning of already familiar animals. He suggests something new, such as

> animal copying with the nucleus taken from one species and the egg in which it was implanted taken from another. Donkey and horse can be mated. Will a donkey nucleus in a horse cell give a donkey or something more like a mule? If it would work we might be able to save some vanishing species by transplanting their cell nuclei into the egg cells of foster species. Is the DNA that carries heredity destroyed immediately when an animal dies? If the meat of wooly mammals locked thousands of years in the arctic ice is still edible, perhaps their DNA is still viable and might be injected, say into elephant egg cells to give baby wooly mammals to the elephant. By some such method, perhaps we might achieve "paleo-reconstruction" of the ancient Mexican corn or mummy wheat, or even the flies that are sometimes found preserved in amber.[97]

Fascinating as these possibilities are, the most provocative application of cloning is with human beings. The thought immediately brings to mind visions of walking into a crowded room and finding that everyone but you looks exactly the same. As bizarre as this may seem, says Dr. Joshua Lederberg, "there is nothing to suggest any particular difficulty about accomplishing this in mammals and man, though it will rightly be admired as a technical tour de force when it is first implemented."[98] Lederberg thinks the big breakthrough will come at any time from zero to fifteen years from now. Dr. Kimball Atwood, professor of microbiology at the University of Illinois, thinks human cloning "could be done now" if there was a crash program. In any event, he is confident that the first human clone will emerge from a laboratory "within a few years."[99] Dr. Watson's predictions are contained in a lengthy discourse entitled "Moving Toward the Clonal Man."

The only real impediment to clonal humanity is a proper method for inserting the nucleus of a body cell into an egg and then stimulating this nucleus to take charge of the cell. The solutions to these problems seem to be within sight. Scientists have learned much about how to implant a nucleus into a foreign cell since the advent of cell fusion. As for turning on the cell's "engine," both J. B. Gurdon (who cloned the African frog) and Ann Janice Brothers of the University of Indiana made separate announcements in 1976 that indicate that the switch to control the reproductive power of the cell is almost known.[100]

But why clone? The answers to that are varied. Each researcher seems to have a different motivation. A sampling of some of the more creative includes:

• *Cloning of historical figures.* Dr. Elof Axel Carlson of UCLA would like to see cloning turned on the dead "to bring back individuals (e.g. historical personalities) of identical genotype." As an example, he says, perhaps there is enough genetic material left "to reconstruct King Tutankhamen from his Egyptian mummy." Carlson notes that "the resemblance of these individuals to ancient photographs and paintings will be startling, but their personalities will be no more like their predecessors than are identical twins to one another."[101] Still, it would no doubt impress a museum visitor considerably to view not only King Tut's gold death mask but a real live copy of its original owner.

• *Cloning for social experimentation.* Joshua Lederberg thinks cloning would allow for some fascinating sociological experiments on how much nurture versus nature affects an individual. With cloning, he hopes, "we can clear up many uncertainties about the interplay of heredity and environment; and students of human nature will not want to waste such opportunities." At the very least, he says, we could enjoy "being able to observe the experiment of discovering whether a second Einstein would outdo the first one."[102] This is a thought vigorously seconded by Dr. Danielli, who feels that "many of the hypotheses the social scientists put forward are extraordinarily difficult to test. Critical testing could be much easier if we had available many genetically identical individuals, who, placed in different environments, would give us a more reliable measurement of the plasticity of human nature."[103]

• *Medical possibilities and the clone.* Lederberg sees no reason why an entire clone can't be grown and kept in storage against the day you have a medical problem—"free exchange of organ transplants with no concern for graft rejection."[104] (If the thought of keeping a clone handy in the garage to pillage for spare parts is gruesome, imagine how it would look to the clone!)

• *The Methuselah clone.* Immortality also enters into any discussion of cloning. Many observers predict that body-cell banks will become more significant in the future than sperm or ova banks. In the event of an untimely accident, one of your body cells could be removed from the bank and a twin of you could be developed. Even if these cell banks are not established, according to Dr. Jean Rostand, a physician could take a tissue fragment from a freshly dead body and proceed to culture it in vitro: "there is no absolute reason why we should not imagine the perfected science of the future as remaking from such a culture a complete person, strictly identical to the one who furnished the principle."[105] Through this method, we truly would become immortals, at least in the biological sense.

• *The ESP clone.* Though no one can be positive yet, the current thinking is that clones will have a high degree of psychic compatibility because of their identical genetic make-up. Some scientists think that extraordinary powers of telepathic communication might exist between clones. If so, Lederberg believes, this ability "might be singularly useful in stress occupations —say a pair of astronauts, or a deep sea diver and his pump tender, or a surgical team."[106] Lederberg even thinks that cloning might ease the generation gap—a father and his clonal son could be on a better wavelength.

Resurrecting George Washington through cloning and obtaining spare organs from a clone are certainly some of the wilder possibilities. A more subtle, and so more ominous, vision presents itself when biologists begin to discuss the "types" of people they deem worthy of being cloned. Nobelist J. B. S. Haldane, one of the first eminent scientists to discuss cloning seriously, has suggested that society would benefit from the cloning of geniuses and other exceptional people. Haldane argues that we now save writings of famous people, so why not save their genotype? The clone, of course, would have to be properly educated if it were to achieve the greatness of its parent twin. Toward that end, Haldane favors a state-supported retirement at the age of fifty-five so that you could then raise and train your own clone to assume its place in society. Haldane also believes that non-geniuses with special physical abilities —dancers, athletes, soldiers—should be cloned, but at an early age, when they are in their prime. More unique physical traits might be added to the clonal list: people with a high pain tolerance, those with night vision, and even dwarfs for use as astronauts, pretailored to the cramped capsule.[107]

We cannot say whom earlier eugenicists like Francis Galton or Harry Laughlin might have cloned, but it seems likely that they would agree with Dr. Bernard Davis of the Harvard Medical School, who favors the cloning of special individuals on the grounds that they "might enormously enhance our culture . . ." Among those he would like to see cloned are men and women who excel "in fields such as mathematics or music, where major achievements are restricted to a few especially gifted people . . ."[108] No one in the cloning field, it seems, is very much interested in reproducing social critics, reformers, or revolutionaries. Like the cloning program envisioned by Aldous Huxley, the copying of individuals in the near future appears destined to be used as a method for maintaining social stability, rather than innovation.

Dr. James Bonner best sums up the eugenics potential of cloning when he forecasts the multiple cloning of selected individuals. "There is nothing to prevent us from taking two body cells from [a] . . . donor and growing two identical twins . . . As a matter of fact, there is nothing to prevent us from taking a thousand. We could grow any desired number of genetically identical people from individuals who have desirable characteristics." This new frontier of cloning, he envisions, will lift Homo sapiens toward "a new super species of human being. It really appears to be within our power—if not today, then in the very near future—to cause our species to develop along any lines we deem desirable."[109]

For students of *Brave New World,* such clonal advocacy conjures up memories of Mustapha Mond and the possibility of "Bokanovskified" eggs and Alphas, Betas, Gammas, and Deltas. Dr. Gunther Stent of the University of Southern California believes that for most people "the idea of cloning humans is morally and aesthetically completely unacceptable." While "it would be fun to have Kant, Beethoven, Isadora Duncan, Einstein, Clark Gable and Marilyn Monroe living on your block," he muses, "the thought of having hundreds of thousands of their replicas in town is a nightmare."[110] Dr. Theodosius Dobzhansky of Rockefeller University agrees: "It can show no lack of respect for the greatness of men like Darwin, Galileo, and Beethoven, to name a few, to say that a world with many millions of Darwins, Galileos, or Beethovens may not be the best possible world."[111]

Dr. Willard Gaylin, of the Institute for Society, Ethics and Life Sciences, is disturbed that problems associated with clonal possibilities aren't taken more seriously by society. Cloning, he says, unfortunately provokes "more titillation than terror, visualized as a garden of Raquel Welches, blooming by the hundred, genetically identical from nipples to fingernails."[112] Gaylin believes our ability to clone represents an "awful potential." Robert T. Francoeur, author of *Utopian Motherhood,* asks, "Xeroxing of people? It shouldn't be done in the labs, even once, with humans."[113]

Bernard Davis, on the other hand, believes concern over cloning has been blown out of all proportion. He agrees that "extension to humans would indeed have grave and novel moral implications," but he argues that "the dangers are hardly terrifying. If human cloning becomes feasible, and if it is then proscribed, an occasional violation would not shake the heavens."[114] Joshua Lederberg, one of the earliest and loudest advocates of cloning, believes that "there is no urgent social problem to be addressed by such a technique . . . Cloning a man is one of the least important questions I can think of."[115]

The United States Congress and members of the legal profession argue differently. A House of Representatives investigation warns that "this technique [cloning] might be further refined to permit the introduction of DNA, wholly synthesized in accordance with some set of specific criteria, and thereby achieve some ultimate objective in genetic engineering."[116]

Though most Americans have probably never heard of clones, civil libertarians and lawyers are already examining their "rights." Lederberg himself has conceded that a new form of racism, "clonism and clonishness," may develop.[117] If clones are looked down upon as little more than carbon copies of life, discrimination is sure to follow. How will society react if a clone of a great man or woman turns out to be a deadbeat? Should the state license cloners to prevent egomaniacs from establishing themselves as a small army?

Bio-law may well become the most important case area in the profession. In a paper entitled "Government Control of Research in Positive Genetics" that appeared in the *University of Michigan Law Review,* I. Scott Bass argues that cloning may make a shambles of our long-held concepts of civil and individual rights:

> Protection of the interests of society as a whole must be considered. Any errors in cloning will be irreversible. The government might be confronted with the problem of arbitrarily classifying the progeny as unsuccessful clones, which are to be killed, or as humanoids, who will be permitted to live at government expense as a reminder of the imperfect operation of cloning techniques.
>
> Incomplete knowledge can impose even greater burdens against which society may wish to legislate. While acquisition of greater knowledge is an integral part of scientific progress, fundamental errors in early cloning techniques might take decades to discover and . . . the cloned individuals would be ruined.[118]

4
eliminating the "bad" genes

Test-tube babies, deep-frozen embryos, and cloning are the exotica of the Genetic Age. Each clashes dramatically with many of our long-held beliefs about procreation, the family, and social relationships. But as methods of eugenic control, none is totally sufficient in itself. The Canadian biologist N. J. Berrill has summed up the limited potential of such techniques: "Artists cannot be counted on to breed artists, nor do astronomers breed astronomers. Nor can the inheritance of general intelligence be predicted. Unusually intelligent parents can produce human vegetables as readily as do other couples, while individuals of exceptional merit tend to crop up everywhere in ordinary run-of-the-mill families."[1]

Ironically, the fusion of genetic exotica with what seems like more "reasonable" techniques of human engineering—genetic screening and genetic surgery—*can* produce the eugenic society. A test-tube baby grown simply from a normal egg and sperm would be a significant oddity; when this process is combined with prenatal screening and gene surgery, the baby can be predesigned for a host of qualities. A clone of a well-proportioned athlete may well be a superior physical specimen, but it is just a shell. If that shell were filled with new genetic information through gene surgery, the result could be a super-athlete, smarter, stronger, faster, and with more stamina than anyone now alive.

Genetic screening and surgery are the product of the

Biological Revolution of the past twenty-five years. At the same time, they represent technological reincarnations of old-fashioned eugenics. Screening, like negative eugenics, aims at eliminating certain genotypes that are deemed undesirable; gene surgery, the new form of positive eugenics, will be used to genetically "upgrade" individuals toward a standard of perfection. As public acceptance of these new technologies grows, which is exactly what is happening at this very moment, so too will the probability that voluntary genetic programs may be legislated into mandatory social-health measures.

Many observers think that that is exactly what will happen. Once the process is begun, they point out, there is really no logical place to stop. If diabetes, sickle cell anemia, and cancer are to be screened for and surgically cured by altering the genetic make-up of an individual, why not proceed to other "problems" of the human condition: myopia, colorblindness, perhaps ridding society of the trait that determines if a baby will become left-handed? As we delve further into the world of the gene, Lawrence Eron of Harvard has warned, we might "phase everybody into the same sort of skin color, height, personality, making it appear as an aid to humanity."[2]

But what color would we select? And what kind of personality? Genetic screening and surgery are founded on the principle that there are certain kinds of genotypes that should be eliminated and others that should be spread as widely through society as possible. But are there any ideal genotypes? Who will make the determination? A black, an Italian, a WASP, and a Chicano might all have different opinions on those questions.

Genetic screening and surgery are likely to touch our lives more deeply than many other aspects of the Genetic Age. After all, only the select few will be cloned, but all of us may one day be screened to determine our genetic fitness. As one observer has noted, none of us can assume we will be immune from this future:

> The time ahead is uncharted. No one has been there, so there are no experts. Each of us whose body and brain may be modified or whose descendant's characteristics may be predetermined has a vast personal stake in the outcome. We can help to insure that good will be done only by looking to it ourselves. We must be careful to retain the individuality of the individual and the personality of the person, or else the humanity of the human may be lost.[3]

SCREENING FOR "BAD" GENES

• In Minneapolis, a married couple sits in the office of a genetic counselor. They are about to be informed of the statistical chances that their next baby will have a genetic deformity.

• A doctor in Los Angeles uses a needle to slowly withdraw amniotic fluid from the womb of a three-months-pregnant woman. Analyzing the chromosomes in the liquid, the doctor will be able to give the patient a good idea of the genetic quality of her unborn child.

• A baby is born in Buffalo, New York. One of its first life experiences is having a few drops of blood pricked from its foot. Doctors will use the blood to screen for the presence of a dozen genetic disorders.

Genetic screening—the testing of individuals for various chromosomal defects—is already an established medical practice.

Few would argue that it is immoral or unethical to spare parents the burden of giving birth to a Down's syndrome (Mongoloid) child, or detecting genetic defects early in a baby's life. But how far do we want to go in analyzing our genetic make-up? The National Institute of General Medical Sciences (NIGMS) points out that twelve million Americans carry true genetic diseases, wholly or partly caused by defective genes or chromosomes. Genetic defects, the agency's research indicates, are present in 5 percent of all live births. Eighty percent of all babies born mentally retarded are the victims of genetic disease. Add to the equation the fact that an estimated two thousand human disorders are genetically determined, and that each of us, no matter how healthy, carries three to eight "bad" genes that could conceivably cause some of these disorders, and the problem begins to take on epic proportions.[4]

That is just what many proponents of genetic screening argue. Only a mandatory, national genetic screening program, they say, can be effective in combating a medical problem of this size. Current laws requiring that children be vaccinated for various diseases are often cited as precedents for legislated screening. Such an undertaking, says the National Research Council of the National Academy of Sciences, would be a progressive health concept.[5]

Dissenters, however, argue that we are entering a field of genetic scrutiny without a firm understanding either of the diseases we are attempting to analyze or of the social implications. An organization concerned with bioethics, the Institute of Society, Ethics and the Life Sciences, has warned that "the potential advent of widespread screening raises new and often unanticipated ethical, psychological and socio-medical problems for which physicians and the public may be unprepared."[6] Marc Lappé fears that if individuals are one day screened for literally hundreds of genetic disorders, as some would like, we might experience "a subtle shift in the way we identify people as humans. We could say, 'Oh, he has an extra chromosome.' We could then identify him as being qualitatively different. He is somewhat imperfect, perhaps not as human as the rest of us."[7] An exasperated physician asks rhetorically, "should gene charts be exchanged before the first kiss, or should there be compulsory premarital screening?"[8]

We are not exchanging gene charts yet, but, technically at least, screening provides enough information to make it possible to do just that. There are currently three types of genetic screening procedures in widespread use, each designed to examine a different phase in the life of a human being: (1) the screening of adults who wish to be prospective parents; (2) fetal screening, or amniocentesis; and (3) postnatal screening of newborn infants.

Just twenty-five years ago, there were no more than a dozen screening and counseling centers around the country; today there are well over two hundred.[9] Ninety percent of the customers are young married couples who have already had a child born with some type of genetic disorder, and are concerned about the possibility of having another. Using family histories, newly developed methods of chromosomal analysis, and the latest theories on the transmission of genetic disorders, genetic-counseling clinics attempt to provide parents with the statistical likelihood of their transmitting a disease to their child.

Genetic disorders are inherited in one of two ways—through either recessive or dominant genes. If one parent has a single "faulty" gene that dominates its normal counterpart, the chances are fifty–fifty that it will be transmitted to the child, and affect it adversely. If both parents carry recessive genes that are passed on, the chances are one in four that the child will inherit the disease. There are also certain genetic disorders that affect only males.

Genetic counselors use this basic knowledge on the transmission of genetic defects to determine a couple's chances of transmitting their disorders to a child. Based on this information, parents may decide to take a chance, or forgo having a child.

Prenatal screening, or amniocentesis, is a much newer technology, though its use is mushrooming. In 1970, estimates are that just two hundred fetal screenings had been attempted.[10] Today, individual doctors have screened as many as seven hundred fetuses apiece.[11] Dr. James Bonner predicts that amniocentesis will "ultimately become standard practice in the developed countries."[12]

The process itself is relatively painless and can be performed on a woman as early as her eleventh week of pregnancy, though the sixteenth or seventeenth week is recommended. A doctor inserts a long needle into the amniotic cavity of the womb and removes some of the amniotic fluid. Because this fluid surrounds the developing fetus, it contains samples of the fetal cells. Through microscopic examination of these cells, lab technicians can detect with near certainty the presence of some seventy genetic disorders. The sex of the baby can also be easily determined. If tests indicate the fetus possesses a disease, the woman's only alternative to bringing it to term is abortion.

Recent techniques for staining and analyzing the chromosomes of the fetal cell will make it possible, in a matter of a few years, to screen for hundreds of genetic traits. By the mid-1980s, a woman may receive detailed knowledge of her offspring a full six months before it is born.

Computers will soon play an important role in fetal screening. Scientists at NASA's Jet Propulsion Laboratory in Pasadena are experimenting with a computer that can analyze a fetal blood sample and then print out a complete chromosomal picture.[13] Dr. Robert S. Ledley of the National Biomedical Research Foundation is another who has developed a fetal screening computer. His device can analyze a chromosome sample in less than forty seconds and compare it against other chromosome charts that have been programmed into the data bank.[14] Many scientists predict that this technique will eventually be used to provide highly detailed analyses of individual genotypes at birth. These charts could be used to eradicate virtually all genetic disease by preventing the mating of couples with the same genetic defects.

The most widespread screening technique of all is postnatal screening, usually performed on babies immediately after birth. Unlike adult screening and amniocentesis, postnatal screening has been the subject of extensive legislation that makes screening for certain genetic diseases mandatory. Today 90 percent of all newborn babies are screened before they leave the hospital, most without the knowledge or consent of their parents.[15]

The first genetic disease for which a screening test was developed was PKU, or phenylketonuria, a disorder caused by the lack of a single enzyme. Untreated, PKU leads to severe mental retardation; 40 percent of those born with the disease have an IQ of less than ten. By detecting PKU very early in life, doctors can prescribe a special diet for the baby. The diet allows the child to develop normally. Today, forty-three states have PKU-screening laws.[16]

The PKU screening test opened the floodgates for postnatal genetic examinations. Throughout the past ten years, as the cause of a disease has become known, a screening test for it has been established. Tay-Sachs, for instance, is a genetic disease whose cause was not discovered until 1969; within two years, large-scale screening had begun in the Baltimore-Washington area. In the words of one doctor, the implementation of screening programs based on rudimentary knowledge is leading to "a disease of the month club" syndrome.[17]

According to Philip Reilly, a legal scholar at the

University of Texas Graduate School of Biomedical Sciences in Houston, fourteen states have mandatory infant-screening programs for diseases other than PKU, and at least twelve more are considering broadening their programs. Since 1974, for example, all infants born in hospitals in New York have been screened for a host of genetic disorders: PKU, sickle cell anemia, maple syrup urine disease, galactosemia, homocystinuria, adenosine deaminase deficiency, and histidinemia. At Boston Hospital for Women, all babies are routinely screened for a number of major genetic defects.[18]

Some politicians are anxious to expand these screening programs by the enactment of federal legislation. In 1975, Congressman Edward I. Koch of New York introduced legislation into the House of Representatives that would extend the New York program nationwide. Senator Hubert Humphrey did the same in the Senate. His bill required all hospitals receiving federal funds to "routinely test newborn infants for metabolic disorders that could retard brain development." Neither bill became law—not because of any opposition to the principle of the program, but because some legislators felt that the nation's hospitals were not yet sufficiently equipped to handle such a task.[19]

No one objects to preventive medicine. What many find troublesome in genetic-screening programs, however, is the unstated assumption that some people have "bad" genes. What is a "bad" gene? Since each of us has an estimated three to eight defective genes, it is conceivable that a 100-percent-effective screening program would show that all of us fail the test of perfection. But not all of us do. The fate of those who are screened in current programs is to carry the burden for us all. "Like it or not," writes Barbara Culliton in *Science* magazine, "there is always the problem of making people feel stigmatized by telling them there is something wrong with their genes."[20]

The sickle cell anemia screening program of the

past decade is a good example of how "gene stigmatization" can develop.

Sickle cell anemia is a disease that affects only black people. Like many diseases resulting from "bad" genes, sickle cell anemia is related to environment. In Western Africa it provides a natural defense against malaria; thus, the natives who carry the "disease" are actually healthier than those who do not, a fact discovered by white colonists.

Sickle cell anemia causes no harm in people who carry it as a recessive trait, but if passed on through dominant inheritance to a child, it proves fatal. There is no cure for those who inherit the disease; they often die very young, before they can transmit their disorder to the next generation.

In the 1960s, sickle cell anemia became a political *cause célèbre.* By 1972, Congress had passed the Sickle-Cell Control Act, providing states with money to launch screening programs. Immediately, significant problems in the program became evident. Screening children was a way to diagnose the disease before it manifested itself, but since there is no cure for the always-fatal disorder, critics questioned the value of informing a child and his parents that he would die young. As Dr. Amitai Etzioni, an opponent of the screening program, protests, "you can't help a seven-year-old, and you make him feel like a freak."[21]

The sickle cell screening program not only does not provide medical help to those affected by the disease, but it has led to widespread confusion about the important difference between those who simply carry the sickle cell *trait* and those who carry the *disease.* Carriers have no health problem; victims die. Because of the confusion about what the screening tests showed, insurance companies have charged blacks higher rates, even though they were not diseased. Some blacks reported losing jobs after being screened. Six airlines made it their policy to refuse to hire blacks who carried the sickle cell trait. The one positive result of the program predicted by its proponents—dissuading people who carried the trait from marrying each other and passing it along to the next generation—never materialized. In recent years, several states have abandoned the sickle cell program.[22]

Many critics see in the sickle cell example problems inherent in much of the screening process: lack of solid medical evidence, social stigma, and psychological consequences. Amniocentesis presents another, and deeper, ethical problem. Of all the genetic technologies currently being developed, this form of prenatal screening offers the greatest benefits to the individual. Through amniocentesis, and subsequent abortion, parents can be spared the agony of giving birth to a baby afflicted with a condition such as Down's syndrome, which would prevent the child from ever being able to lead a normal life in society. While strict anti-abortionists might reject even this, most people would probably sympathize with a woman's decision in such a case. But when such a process is magnified throughout society, other considerations begin to enter in. Paul Ramsey, the well-known ethicist, believes a screening technique like amniocentesis "seems destined to degrade society's willingness to accept and care for abnormal children, and at the same time, to enlarge the category of unacceptable abnormality, while narrowing the range of acceptable normality."[23] Leon Kass fears we are setting the stage for a social attitude that will be directed against a child with genetic defects, as someone "who need not have been, and who would not have been if only someone had gotten to him in time."[24] Their fears are given substance by Dr. Cecil B. Jacobson, chief of the Reproductive Genetics Unit at George Washington University. Jacobson, one of the initial developers of the technique of amniocentesis, proposes that we broaden its application to prevent a wide range of "abnormalities":

> The mongol question is just the most superficial part of the whole problem. But I'd like to take the thing a few steps further. Would you, for instance, want to conceive a child who will die of cancer at age 40, if the tendency for the development of cancer can be shown before birth? Naturally, at this point we're not able to do that. But if we could tell what fetuses are going to be affected with cancer in their 40s and 50s, I would be for aborting them now. That would eliminate some types of cancer forever.

Jacobson says he would also favor abortion "if both parents want only girls and the current pregnancy tests by amniocentesis to be a boy. I just don't recognize any absolute here."[25]

Jacobson's strong advocacy of aborting fetuses for any of a number of reasons, both medical and non-medical, points out the critical nature of the relationship between the patient and the genetic screener. Patients often don't understand the fundamentals of genetic disease. A study conducted at a Philadelphia hospital, for instance, showed that after extensive education and counseling about a specific genetic disorder, only 19 percent of the patients could grasp the medical nature of the disease in question.[26] Because of the complexity of genetic disorders, patients must rely heavily on the guidance of their counselor. But to whom does the counselor owe his primary obligation? Many doctors who perform screening tests, like Dr. Jacobson, feel they have a major duty to prevent genetic disease in future generations by guiding parents to abort their own "defective" babies. Marc Lappé worries that counselors are sacrificing their patients' best interests in the name of purifying the gene pool of the next generation. "To maintain that we are adding to the public good by conscientious genetic guidance and screening may be fallacious and ethically dangerous. The genetic counselor's obligations should never extend beyond the family within his purview." Dr. Margery Shaw, director of the Medical Genetics Center at Houston, believes the question which remains unanswered is "do our genes belong to us, or do they belong to future generations?"[27]

John Fletcher of the Hastings Institute (not to be confused with Joseph Fletcher of the University of Virginia) wonders if parents aren't being given more information about themselves and their children than they are prepared to deal with. "There is often an unusual sense of shame and guilt associated with genetic disease," he points out. "The tendency of parents and doctors to reject the genetically defective is an all-too-natural one. I don't want to see amniocentesis and applied genetics destroying our capacity for support and compassion for those who don't measure up to our norms."[28]

The rapid growth in the field of screening is setting the climate for a revitalized negative eugenics, where "defectives" (defined in whatever way society determines) will be disposed of like the unwanted babies of ancient Sparta. Dr. Bentley Glass enthusiastically hopes that "genetic clinics will be constructed in which before long, as many as 100 different hereditary defects can be detected in the carriers, who may be warned against or prohibited from having offspring."[29] Another biologist remarked to a human-genetics conference several years ago that "I have visions of a future genetic clinic in which a person will have not one, but hundreds, of his proteins analyzed completely in short order. The results will be run through a computer, and a license to reproduce will then be issued on the basis of a passing grade with respect to his (or her) genes." He concluded his remarks by adding, "as my vision begins to assume the proportions of a nightmare, I shall abstain from further speculation along these lines."[30]

Two members of the Hastings Institute, Bruce Hilton and Daniel Callahan, have both warned that screening for "defectives" may inevitably lead to mandatory regulation of the quality of genotype necessary for birth. Callahan observes that the very act of providing expectant mothers with genetic information about their fetus allows them to decide whether to keep or abort the baby. "Giving people freedom of choice," he says, "is to make them responsible for the choices they make. It is then only a very short step to begin distinguishing between responsible and irresponsible choices, with social pressure beginning to put in an appearance." Callahan predicts that in the future parents who choose to keep defective children will be looked upon as irresponsible, especially in light of the potential social expense required to keep, say, a Down's syndrome baby alive. When cost analysis enters into the question, he suggests, we will finally arrive at the point of putting a "price on everyone's head."[31]

Hilton foresees a day when public opinion determines that it is in the interest of society "to abort certain fetuses whether we want to or not." What will happen to our social framework, he asks, "when civil liberties increasingly conflict with public health priorities?"[32]

Dr. Jon King believes that screening is creating a preoccupation with "genetic defectives," and he wonders where it will end. "None of us would call blue eyes a genetic disease," he notes, "although blue-eyed individuals differ genetically from brown-eyed ones." Clearly we would not stigmatize those people having blue eyes as abnormal, nor would we advocate that blue-eyed fetuses be aborted. Yet, King insists, "many genetic screening programs are based on just this idea, reinforcing the myth that genetic variation is equivalent to genetic abnormality."[33]

Daniel Callahan agrees, and points to a disturbing unstated eugenics philosophy that may be central to programs that are designed to screen out the genetically abnormal: "Behind the human horror at genetic defectiveness lurks, one must suppose, an image of the perfect human being. The very language of 'defect,' 'abnormality,' 'disease,' and 'risk,' presupposes such an image, a kind of proto-type of perfection."[34]

GENE SURGERY

Genetic screening is a detection mechanism; it allows researchers to pinpoint couples likely to transmit genetic diseases, anticipate genetically defective fetuses, and analyze newborns for genetic disorders. On the basis of information received from screening, couples decide to refrain from having children; fetuses are aborted; newborns may be treated with special diets or medicines.

Genetic surgery goes one step further, representing the final solution to the problems detected through screening. By providing the tools to manipulate genes and chromosomes through the introduction of new genetic information into the cells, genetic surgery is the single most powerful technique of genetic engineering. Cloning and test-tube babies may provoke a more squeamish response from most of us, but genetic intervention—through the use of recombinant DNA, gene synthesis, cell fusion, etc.—embodies the ultimate in biological control. Dr. Salvador Luria of Harvard issues a chilling warning when he says, "we should not ignore the possibility that genetic means of controlling human heredity will be put to massive uses of human degradation . . ."[35]

As a technique of the genetic engineers, gene surgery is truly in its infancy. The first reported instance of its use in an animal came in 1959, when three French scientists claimed that they had taken DNA from one type of duck, Campbells, and injected it into a second breed, Pekins. Their experiment was designed to influence the genetic composition of the Pekins' offspring, but to their surprise, the injected duck began to change markedly, taking on some of the physical characteristics of the DNA donor. This development created a wave of interest, causing scientists around the world to attempt to duplicate the results, all unsuccessfully.[36]

The fruit fly has proven more amenable to repeated experimentation. Two researchers, A. S. Fox and S. B. Yoon, have devoted their professional careers to the study of the fruit fly, *Drosophila.* Using two different strains of the fly, Fox and Yoon have managed to incorporate DNA from a mutant type into a normal *Drosophila.* Several of the resulting offspring took on some of the characteristics of the mutant, indicating that the original DNA had been transformed.[37] A Ukrainian scientist, Dr. Serge Gerhenson, claims to have performed a similar experiment using a silkworm.[38] More recently, the heredity of mice has been changed permanently through genetic surgery.[39]

There is only one known, and still extremely controversial, case of human genetic surgery. In 1970, two German girls, ages seven and two, were brought to a hospital. Suffering from a rare genetic disorder caused by an enzyme deficiency, the girls exhibited palsy, epilepsy, and severe mental retardation. German doctors, working with Dr. Stanfield Rogers of the Oak Ridge National Laboratory, injected a virus into the girls that carried the genetic information they hoped would correct the deficiency. In 1975, Dr. Rogers reported that the attempt had been a failure, possibly because the virus had been inadequately stored before use. Many doctors have protested the surgical effort, because it is known that the virus used on the sisters causes cancer in test rabbits. Dr. Rogers, however, has indicated that he will attempt further genetic surgery on human patients in the future.[40]

In May, 1977, Drs. William Rutter and Howard Goodman of the University of California at San Francisco announced a major research breakthrough that may make Dr. Rogers's next try at genetic surgery a success. The California scientists reported that they had successfully used a recombinant DNA to turn *E. coli* bacteria into a cellular "factory" that manufactures the genes necessary to produce insulin, a drug required by diabetics. Their next step is to "switch on" these genes to begin insulin production, something Dr. Rutter says will be done "sooner than anyone has expected."[41] The final hurdle to be cleared before genetic surgery can be used on diabetics is the discovery of a method to introduce the insulin genes into body cells, where they can express themselves chemically.

This very work is a high priority of government-funded medical research. The National Institute of General Medical Sciences yearly awards over $117,000,000 to researchers concentrating on developing methods of genetic surgery. According to NIGMS, the thrust of the research conducted by its grantees "focuses on . . . the mechanism of transmission and expression of genetic information."[42]

The government currently supports ten national centers for medical genetic research in major cities, including Houston, Philadelphia, New York, and Seattle. In addition, NIGMS makes funds available to over three hundred smaller projects that explore various aspects of genetic disease. The centers, which bring together a core of geneticists surrounded by staffs of biochemists, molecular biologists, physicists, and other specialists, are the backbone of the government's genetic-research effort. Through them, federal health authorities hope to arrive at an understanding of rudimentary genetic disorders that will lead to the eventual treatment of more complex genetic functions.[43]

The NIGMS publicly predicts that these centers—which are researching the genetic causes and cures of such varied problems as cancer, diabetes, and myopia (nearsightedness)—will provide important results: "Clearly, the pay-off for research dollars invested promises ultimately to be higher than that expended for practically any other line of medically oriented research, so sweeping is the horizon of genetic medicine for combatting disease."[44]

According to NIGMS doctors, all of us eventually fall susceptible to genetic disease. One of NIGMS's principle grantees, Dr. C. J. Epstein of UC-San Francisco, argues that, "carried to the *extreme* virtually all of human disease and deterioration can be considered as birth defects, since genetic disorders are undoubtedly of importance in their origin and pathogenesis."[45] Dr. Vasken Aposhian, formerly of the University of Maryland, believes that once he and his colleagues "can deliver the DNA that could cure or treat" this myriad of diseases, they will be able to genetically administer cures "over the entire span of lifetime of various individuals."[46]

Aposhian and others plan to use their genetic surgical skills to replace defective genes, introduce new genetic material, and activate genes that lie masked and dormant. Just a few years ago, notes Dr. Theodore Friedmann of the School of Medicine, University of California at San Diego, "examinations of the prospect for gene therapy for human genetic disease took a certain comfort from the degree of remoteness and unreality that characterized the subject."[47] Dr. Aposhian, for one, believes that "it is now conceivable that not within 100 years, not within twenty-five years, but perhaps within the next five to ten years, certain inborn errors of metabolism will be treated or cured by the administration of the particular gene that is lacking."[48] Even at that, Aposhian may be too conservative in his countdown to genetic engineering. The announcements from UC-San Francisco in May of 1977 show that recombinant DNA is adding new velocity to the gene surgeon's pursuit.

Initially, genetic surgery will be turned on correcting or changing monogenic traits, those simplest to handle.

Most researchers agree that diabetes will be operated on first. If everything works according to plan, new genes will be introduced into the cell, and diabetics will produce their own insulin, without the necessity of daily shots. "At this point," Dr. Robert Sinsheimer of Cal Tech believes, ". . . it is clear that technically and literally the stage is set" for applying genetic engineering to the problem of diabetes.[49] Other much-discussed possibilities include genetic cures for hemophilia, PKU, Tay-Sachs, and, many hope, cancer. The enzyme deficiency or genetic defect is already known for ninety-two different monogenic diseases.[50] Presumably, the curing of just one will provide basic knowledge on gene expression that will apply to all. The future of the technology is almost limitless, suggests Aposhian: "if one considers the purpose of a drug to be to restore the normal function of some particular process in the body, then DNA should be considered to be the ultimate drug."[51]

Some researchers are not so certain that Aposhian's ultimate drug will produce the harvest of good health he predicts. Critics of genetic surgery point out that our technical capability may permit the injection of new genetic material into someone, but our knowledge of the biological side effect of such a move is close to nil. Drs. Friedmann and Richard Roblin note that "any potential large-scale use of gene therapy (for example, the prospect of treating the approximately 4 million diabetics in the U.S. with DNA containing the gene for insulin) might appreciably affect the overall quality of the gene pool and would require other forms of control." We might, they feel, cause a larger medical problem than the original one posed. They argue that "the irreversible and heritable nature of gene therapy means that the tolerable margin of risk and uncertainty in such therapy is much lower than with other experimental medical treatments."[52]

Others worry that gene surgery will lead to a narrowing of the gene pool as we weed out "bad" genes. Since part of the strength of our gene pool consists in its very diversity, including defective genes, tampering with it might ultimately lead to extinction. Critics recall that in the 1950s genetic modifications were made in wheat strains to create bumper crops of "super-wheat." When a new strain of disease hit the fields, farmers found that their wheat was too delicate to resist. Within two years, virtually the entire crop was destroyed.[53]

If we are to be redirected and modified, will we become susceptible to a new plague, too? Marc Lappé cautions that we might: "Genetic diversity is in one sense capital for investment in future adaptations. Since genetic variability represents evolutionary capability, it is a load we should be ready and willing to bear."[54]

Assuming that we do not go the way of the 1950s crop of super-wheat, a more insidious danger awaits us as genetic surgery gains in sophistication. As knowledge about monogenic traits increases, the bioengineers will inevitably gain new insights into the functioning of more complex characteristics, such as those associated with behavior and thought. As justification for extending genetic surgery to the "redirecting" of human emotion and intellect, many molecular biologists are already contending that schizophrenia and other "abnormal" psychological states result from genetic disorders. Others now argue that "antisocial" behavior, such as criminality and social protest, are also examples of malfunctioning genetic information. Dr. Ralph W. Gerard, a prominent neurophysiologist, has put the case most bluntly. "There can be no twisted thought without a twisted molecule."[55] Sociobiologists go even further, claiming that virtually all human activity results from genetic determinism that makes us act in accordance with our remote evolutionary past. If we wish to change this situation, it is claimed, we must change our genes. As such thinking gains credence, and recombinant DNA and other methods of gene surgery are refined, it is only a matter of time until "the ultimate drug" is turned into "the ultimate psychiatrist" and finally "the ultimate brain modifier."

Lord Ritchie-Calder believes that if genetic engineering continues on its present course, behavior modification through gene surgery is not only inevitable, but the only plausible conclusion of the Genetic Age. "Since we no longer count the stopping of respiration or the heartbeat as the sign of death," observes Ritchie-Calder, "but rather the cessation of brain waves as revealed by the encephalograph, the logic of all this bio-engineering is to keep our biological computer, the brain, going."[56] The question, of course, is where will it be going? And who will be in the driver's seat?

Genetic engineers are not the first scientists to seek control of the brain and its functions. The "Psychoneurological Revolution," stepchild to the Biological Revolution, is already in full force. If the brain research that has been undertaken to date is any indication, there can be little doubt that the use of genetic surgery to transform emotion and behavior will not usher in a new era of human liberation.

As though in a scene from the novel *Clockwork Orange,* a myriad of drug experts, electrical wizards, and behavior-modification specialists try their hand at controlling our gray matter. Inserting electrical implants at different points of the brain, researchers can make a charging bull stop in its tracks and wander off aimlessly. An implanted human can be turned into an epileptic in the midst of a violent attack, or provoked to sustained laughter for no apparent reason at all. Commenting on the use of electric stimulation to induce laughter and happiness, Sir Julian Huxley admits that "to some people, this seems somehow too materialistic; but after all, electric happiness is still happiness, and happiness is very much more important than the physical happenings with which it is correlated."[57]

Electrical happiness may indeed still be happiness, but, as bioengineers are fond of pointing out, brain control through electricity, chemicals, and behavior modification suffers from one glaring defect: It is not permanent. An electrode, after all, must be switched on and off; a drug must be administered regularly. Genetic change is the only permanent change.

Interest in neurobiological research is a relatively new phenomenon among molecular biologists. But now that most of the great puzzles of gene structure and function have been solved, top researchers in the field are increasingly turning to the mysteries surrounding the central nervous system and the brain. Dr. Marshall W. Nirenberg of NIH and Dr. Max Delbruck of Cal Tech, both Nobelists, are already well entrenched in nerve research.[58]

How far off is brain modification through genetic engineering? While the chemical nature of emotions like hate, love, happiness, and aggression are still not understood, W. French Anderson believes that with our present state of knowledge, it is at least possible to produce "defective intelligence (mental retardation) by any one of many different single gene mutations."[59]

More subtle and significant forms of biocontrol remain to be developed in the future, but in a field where knowledge is doubling every two years, that future may not be so very far off. Extensive research has already been conducted on the role that RNA plays on memory and thought. Experiments conducted on animals have produced methods for destroying memory through RNA manipulation. Researchers have also learned how to transfer memory and intellect from one animal to another by using RNA injections.

While it is unlikely that genetic manipulation of the human brain will lead to complete control of all specific behavior and thought patterns, it seems quite possible that broad aspects of human activity can be biologically altered à la *Brave New World*. As Harvard's Bernard Davis concedes, "we cannot exclude the possibility that a few key genes might play an especially large role in determining various intellectual or artistic potentials or emotional patterns."[60] All experts agree that fetal engineering, or the introduction of new genetic material into the sperm or egg, would be a much simpler method of manufacturing preprogrammed people.

Still, who among us is wise enough to determine genetic alterations involving intelligence, compassion, tolerance, anger? Unnecessary aggression is condemned by our society, but should we exterminate it with genetic surgery if we discover its biochemical basis? Are the happiness and self-satisfaction of *Brave New World* the fulfillment of human destiny?

Charles Frankel, professor of Philosophy and Public Affairs at Columbia University, questions our sophistication in undertaking the task of recasting ourselves in our own images of what we think we *should* be. If the medical and evolutionary considerations make genetic engineering inadvisable, he says,

> the facts about human desires make it doubly so. For people are probably wisest, they understand their desires best, when they are aware that they do not know what they desire human beings to be forever and ever. In time of war, the qualities sought are resilience, discipline, loyalty, physical courage; in time of peace, people praise flexibility, independence, moral courage. Assuming that we had the requisite powers, shall moods and needs of a decade or half-decade be permitted to affect the permanent temperament of mankind?[61]

Dr. Joshua Lederberg makes the counterpoint to Frankel: "It makes as little sense to decry genetic adaptations to medicine as to fire, shelter, clothing or other components of civilized life. The quality of a genotype cannot be evaluated in terms of a hypothetical state of nature, wherein we would quickly grunt in chilly displeasure at our unfurred skins. Rather, it must be able to cope with the changing milieu."[62]

Genes, of course, are not like the trappings of civilization. Genes are internal to us, and when we redirect them, we redefine ourselves.

BEYOND OURSELVES, WHAT?

> Modern genetics is on the verge of some truly fantastic ways of "improving" the human race, . . . but in what direction?[63]

From the scientific viewpoint, each bit of the accumulated molecular knowledge of the past quarter century has given proof to one genetic truth: Life is simply an arrangement of chemicals and biological processes. Nothing more, nothing less. Switch a bit of information here, introduce some new material there, and the chemicals are rearranged, different functions are produced. It is this truth of molecular biology that opens the way conceptually and philosophically for the human species to move beyond itself. Since there is nothing "sacred" about our molecular arrangement, some argue, why content ourselves with simply repairing defective genes when we can create entire new patterns and forms of life?

Dr. Joshua Lederberg believes we will enter a posthuman age because "it is possible to design a useful protein from first premises, replacing evolution by art."[64] Since "the present genetic types of man may not all permit a happy adjustment" to the world of the future, Dr. Bentley Glass argues that "the nature and personality of man must change . . ."[65] Dr. Shapley says that "man as half beast, half angel, must of course comply with the biogenic law, but he is able to make amendments thereto." Shapley looks toward a day when geneticists "will correct our mental and social structure."[66]

Dr. Francis Crick believes that as this new world unfolds, the wilder genetic scenarios will not be instituted because "people will simply not stand for them."[67] Many would argue that point, based on our track record of the past two decades. Dr. Jon King senses there is a gathering momentum of self-justification within molecular biology that "has spawned technologies in need of socially useful applications in the same way that the space program has generated specialized engineering technologies now searching for some social use."[68]

But what kind of "socially useful applications" are imagined? We have already mentioned various new varieties of food products, medical treatments, and predesigned animals. There is another realm of genetic engineering, however, that few of us can even grasp intellectually because it is so foreign to everything we know. Subconsciously we turn away from the specter of mass bioengineering, pretending that genetic surgery and recombinant DNA will be used solely to make minor medical alterations in human beings. More profound genetic changes are relegated to the ranks of science fiction, even though their technical realization is near at hand.

In May of 1977, columnist and social satirist Russell Baker provided millions of Americans with a few good laughs in a piece about a mad geneticist named Dr. Irving Slezak. Using recombinant DNA, Baker wrote, Slezak combined genes from different organisms to produce all kinds of outrageous chimeras. On one occasion, the scientist linked a gene belonging to a truck driver to another from a state policeman. By blending the two, Baker noted, Slezak "hoped to produce a brand new form of life—a truck driver who, immediately upon exceeding the 55 miles an hour speed limit, would pull himself over and give himself a ticket." The doctor was also intent upon crossing a go-go girl with a seal to produce a topless dancer that could perform in Alaska during the winter.[69]

As outrageous and impossible as these spoofs are, they are hardly more imaginative than the scenarios conjured up by the bioengineers themselves. So extreme are the possibilities that it is impossible to place them within any familiar conceptual framework. It is enough to allow the scientists to speak for themselves. To begin with, consider the body, a favorite clay of the biosculptors.

Nobelist Jean Rostand went on record as believing that "it would be no more than a game for the 'man-farming biologist' to change the subject's sex, the colour of his eyes, the general proportions of body and limbs and perhaps the facial features."[70]

Robert O. Becker of the V.A. Hospital in Syracuse has conducted extensive research into the regeneration of limbs, a phenomenon that occurs with regularity among certain amphibians. Becker has succeeded in stimulating partial regeneration in limbs of rats.[71] Dr. Marcus Singer of Western Reserve University has conducted similar research with frogs. "We cannot rule out the possibility that some day human beings might be able to regrow organs and tissues which they presently cannot."[72] Bonner sees no reason to doubt that: "we should be able to take a single cell and reset the genetic program to any desired point to make that cell or group of cells turn into a new organ, a replacement organ or a new liver. Maybe you will go to the doctor and he will say: 'Well, I think your heart isn't so good now. Maybe we had better start growing you a new one, and in two or three years, it will be grown up and we can plumb it in.' "[73]

In fact, the same technique could be used to produce *extra* limbs on a person—spare legs for long-distance runners, say, or perhaps eyes in the back of the head. For himself, Bonner says, "I have tried to think about what further organs I would like to have, and I have decided that I would like to have four hands since there is so much for biologists to do." With a touch of humor he adds, "Recently, as I was trying to light my pipe in the laboratory, my colleague, . . . said to me, 'If you're going to smoke a pipe in the laboratory, you'll need five.' "[74]

Joking aside, NASA's director of Biotechnology and Human Research, Dr. Eugene B. Konecci, is quite serious when he says, "Our understanding of the chemistry of the genes may soon enable us to modify or enlarge organs, thus accelerating evolution to a state where man could successfully survive in strange environments."[75]

Traditional sexual roles and behavior, of course, are also open to genetic modification, though, as Lederberg notes, "what finesse it will take to design genotypes optimized for both sexes, so that the development of either sex is consistently rechanneled to the full set of desirable qualities, in addition to the primary sex characteristic!" The question, he says, is, should each sex be independently directed, or "shall we bypass the dimorphism and evolve a race where this does not matter?"[76]

Joseph Fletcher's idea of the future of sexuality is more concrete. He foresees a genetically and medically defined unisex:

> . . . transplant or replacement medicine foresees the day, after the automatic rejection of alien tissue is overcome, when a uterus can be implanted in a human male's body—his abdomen has spaces—and gestation started by artificial fertilization and egg transfer. Hypogonadism could be used to stimulate milk from the man's rudimentary breasts—men too have mammary glands. If surgery could

> not construct a cervical canal the delivery could be effected by a Caesarean section and the male or transsexualized mother could nurse his own baby.[77]

Another favorite topic of discussion among the bio-engineers is growth modification of the brain. Experiments are already proceeding. A simple injection of a growth hormone into a rat fetus will cause the brain size to grow 76 percent with equivalent increase in learning ability. Most researchers, however, put their faith in a more subtle genetic manipulation. During development, the brain divides thirty-three times until it reaches its full size of nine billion cells. If just one more division could be artificially induced, the resulting brain size would be enormous. Lederberg, for one, believes it would be incredible "if we did not soon have the basis of developmental engineering techniques to regulate, for example, the size of the human brain by prenatal or early post-natal intervention."[78] The only roadblock he foresees is an anatomical one: The female pelvis is simply too small to permit such a large head to pass through it during birth. The solution will no doubt lie with widespread use of Caesarean sections.

Another school of thinking holds that we would do better in life if we weren't limited to just two biological parents. Why not branch out—say, to four? Dr. Beatrice Mintz has shown that this is possible by taking two mouse embryos from different parents—one type with black coat, the other with white—and fusing the two cells together. She then inserted the combined embryo in a female mouse, which carried it to term. The baby was born normal in every respect but one: It had black and white stripes.[79] Mintz's mosaic mice, as they are called, have inspired Dr. James Danielli to reflect on mosaic people: "One can see that formation of allophenic man, obtained by fusion, at an early age, of two or more complete embryos, may be simple to do and successful. Such an individual would have four or more genetic parents, and hopefully, would have the genetic advantages of all parents." Danielli concedes, however, that there is the unfortunate possibility that his allophenic man might also inherit the "deficiencies of all the parents." In a related possibility, he believes that if the problem were properly tackled "within 10 years, and probably within 5 years, human egg cells could be artificially assembled from nuclei, membranes and other components . . ."[80]

Other changes in life as we know it are also possible. The DNA chimeras may show the way for full-organism chimeras in the future—real-life minotaurs, perhaps, or satyrs. Dr. J. B. S. Haldane has suggested that astronauts could be genetically altered to assume some characteristics of monkeys, complete with tails, long arms, and stooped posture. Why? Because, says Haldane, "a gibbon is better adapted than a man for life in a low gravitational field such as that of a space ship, an asteroid, or perhaps even the moon. Gene grafting may make it possible to incorporate such features into human stock."[81]

Dr. Kimball Atwood, chairman of the Department of Microbiology at the University of Illinois, sees no technical reason why we couldn't eventually "produce an organism that combines the happy qualities of animals and plants, such as one with a large brain so that it can indulge in philosophy and also a photosynthetic area on its back so that it would not have to eat."[82]

The possibility of introducing intelligence into lower life forms raises thorny questions. While Lederberg believes that "before long we are bound to hear of tests of the affect of dosage of human 21st chromosome on the development of the brain of the mouse or gorilla," he wonders, "What are the legal rights of a chimera?"[83] Two lawyers, Charles Weigel and Stephen Tinkler have already considered the matter. Acknowledging that "the introduction of elements of the human genetic constitution into higher order Primates is realizable," they wonder: "Should the law provide some arbitrary ratio which states that all 'creations' that have less than 1/100th human element may be destroyed if in the opinion of the experimenter they are intolerable or even unsatisfactory? Or perhaps 1/10 would be allowable. Or less than one half."[84]

Most of us, of course, have never given much thought to the legal rights of monkey-men, because we never knew their existence was a real possibility. Nor would it occur to the vast majority of people that there is some "advantage" for ourselves and our children to be had by manufacturing babies with giant heads, or hybrids that combine plant and human qualities, or fusing together two embryos to form one baby. Professor Charles Frankel has eloquently observed of those who seek these eugenic changes in humanity: "They seem not to hear themselves. It is that other music they hear, the music that says that there shall be nothing random in the world, nothing independent, nothing moved by its own vitality, nothing out of keeping with some Idea: even our children must be not our progeny but our creation."[85]

Geneticists may protest that the biological futures they conjur up are simply their version of shop talk, fascinating intellectual game-playing. Nevertheless, insists bioethicist Marc Lappé, "the general *motivation* for proposing cloning or other engineering of man must be taken seriously, because it reveals a tacit approval by some of the best minds of the country for both the legitimacy and the need for introducing genetic controls . . ."[86] The fact that researchers like Lederberg, Atwood, Danielli, and others are not science-fiction writers, but rather skilled technicians who already possess many of the tools necessary to institute some of the most bizarre genetic possibilities, must also be taken into account.

To Dr. George Beadle, a Nobelist, these scenarios

are not fantasy but too close to reality. "Man knows enough," he says, "but is not yet wise enough to make man."[87]

Dr. Leon Kass provides eloquent testimony to the enormity of what is at stake:

> We have paid some high prices for the technological conquest of nature, but none perhaps so high as the intellectual and spiritual costs of seeing nature as mere material for our manipulation, exploitation and transformation. With the powers for biological engineering now gathering, there will be splendid new opportunities for a similar degradation of our view of man. Indeed, we are already witnessing the erosion of our idea of man as something splendid or divine, as a creature with freedom and dignity. And clearly, if we come to see ourselves as meat, then meat we shall become. The new technologies for human engineering may well be "the transition to a wholly new path of evolution." They may, therefore, mark the end of *human* life as we and all other humans know it. It is possible that the non-human life that may take our place will in some sense be superior—though I personally think it most unlikely, and certainly not demonstrable. In either case, we are ourselves human beings; therefore, it is proper for us to have a proprietary interest in our survival, and in our survival as human beings. This is a difficult enough task without having to confront the prospect of a utopian, constant remaking of our biological nature with all-powerful means but with no end in view.[88]

5
bio-futures

When scientists talk about cloning a superior breed of humans or developing hybrid species—for example, crossing monkeys and humans to perform certain tasks in society—most people are likely to respond by saying, it can never happen here. In fact, a public-opinion poll done by Louis Harris several years ago found that a majority in this country would be opposed to genetic engineering.[1]

But clones and hybrids will not suddenly be forced upon us by some form of biological *coup d'état*. No such dramatic conspiracy is in the offing. Instead, we can expect a much more mundane step-by-step introduction of new genetic technologies, each seemingly practical in the immediate benefits it can bestow, and collectively providing a seductive if not irresistible framework for the gradual reorientation of human values and the ultimate acceptance of mass genetic engineering.

Part of the potential appeal of bioengineering lies in the fact that our society is fast losing confidence in its ability to shape the external environment to better the human condition, especially as newer technologies heaped on top of older ones create even more unresolved social, economic, and political problems.

Consider the fact that our nation has now sunk to eleventh among the nations of the world in infant mortality, life expectancy, and literacy, and the downward trend is continuing. Pollution of the environment is becoming so acute that millions of workers are subjected daily to the hazards of serious illness and possible death. The problems in adjusting to the stresses of modern urban life—overcrowding, unemployment, inflation, and unresponsive political and economic bureaucracies—are all reflected in the dramatic rise in mental illness in the general population. Finally, the paranoia born of thirty years of struggle to maintain ourselves as the world's preeminent military power and the recent loss of the Vietnam War have shaken the already fragile sense of security that has become such an obsessive element of our national character.

The old approaches to solving America's pressing problems seem no longer to work. At the same time, it's a sure bet that those in power will not ask us to seriously examine some of the underlying institutional causes of the predicament we find ourselves in as a society. Instead, our public leaders now tell us that we will have to learn to live with the escalating crisis that surrounds us and begin adjusting ourselves to a changing external world that we can no longer completely control. In fact, the public is already well on the road to being convinced that it will be too painful or too expensive to make the kinds of drastic changes in our social and economic environment that are necessary.

It is within this broad context that genetic engineering has a better than good chance of gaining support. For if the manufactured environment is no longer subject to human intervention, then the other option is to begin changing, for the first time, human biology itself to accommodate the new reality in the outside world.

The necessary conditions for this "biological accommodation" already exist in a broad range of categories. Consequently, new genetic technologies are likely to be applied in major areas of American life as soon as they become available.

POPULATION CONTROL AND FAMILY PLANNING

Population-control advocates point out that while there are nearly four billion people in the world today, it took from the beginning of human history until 1830 to produce the first billion, a century more for the second billion, thirty years for the third billion, and a mere fifteen years to produce the last billion.[2]

These statistics become even more ominous in light of studies that show that over one quarter of the world's population is suffering from malnutrition or starvation. Many experts expect this situation to become more critical by the year 2000 as the increased population begins to literally overrun major urban areas and further exacerbate problems of hunger and social unrest. Up to now, population-control programs have focused on decreasing the numbers of births in the population. Now, with geneticists and social planners worrying out loud about the deterioration of the human gene pool, there is a new interest in limiting the types and quality of births as well. Bentley Glass, former president of the American Academy for the Advancement of Science, warns that "in an overpopulated world it can no longer be affirmed that the right of the man and woman to reproduce as they see fit is inviolate . . ." Professor Glass believes that the right of parents to procreate must become a secondary consideration to the "right of every child to be born with a sound physical and mental constitution, based on a sound genotype."[3] Others, like Joseph Fletcher, argue that selective breeding is our only option if we are to survive as a society. The alternative, says Fletcher, is "the enormous cost in medicine, surgery, artificial aids and diet controls for the increased number of victims"[4] which would result from continuing to allow those with bad genes to reproduce.

For those who might be somewhat squeamish about the prospect of selective breeding, Fletcher answers by saying that not to initiate such policies would be immoral and irresponsible, since we have a sacred duty to control our heredity just as we accept the responsibility of controlling our social life and our behavior. Echoing the catechism of earlier eugenicists, Dr. James Bonner of Cal Tech asserts that selective breeding of children is really just an extension of our long practice of domesticating animals and plants, and therefore should not be greeted with any great sense of alarm.

As was pointed out in the last chapter, selective breeding of children and genetic engineering are already practiced through adult screening programs in almost every state, and through amniocentesis and postnatal screening programs in hospitals all over the country. Tens of thousands of couples make decisions every year about the quality of their offspring through these various medical procedures. Presently those decisions are being made, says Fletcher and others, in a haphazard way, often without the knowledge and benefit that a coordinated national program could bring to bear.

Geneticists like Nobel Prize winner Francis Crick agree. In fact, Crick suggests that "it would not be very difficult for a government to put something in our food so that nobody could have children. Then possibly they could provide another chemical that would reverse the first, and only people licensed to bear children would be given the second chemical."[5] While at first such proposals are likely to sound a bit unorthodox to most prospective parents, proponents point out that such a compulsory licensing scheme is no more absurd than requiring a driver's license for those who wish to operate an automobile. Just as society has made a collective decision not to allow just anyone to drive a car, says Joseph Fletcher, it has an equal obligation not to allow just anyone to be "free to go producing children of any old kind they want or happen to conceive."[6]

A good many people are apt to balk at first at the suggestion of a federal registry and licensing for having children. But few, if any, would reject the nction of improved genetic procedures to insure the good health of their offspring. The paradox is that as the scientific and medical knowledge of genetic engineering increases, the prospect of predetermining everything from eye color and sex to polygenic traits like memory retention increases as well. For example, if parents were told by their doctor that they could choose the sex of their child and make sure that it would grow up to be slightly taller and smarter than the norm, wouldn't they feel an obligation to provide it with these options if they felt it would enhance their child's ability to better cope with life? On the other hand, what if millions of parents decided to have boys who would grow up to be taller and smarter than the average? After a decade or so of this kind of continued one-upsmanship the social consequences of runaway biological inflation could be catastrophic. For this reason, some scientists believe that such genetic-engineering decisions would eventually have to become at least regulated by public authorities, rather than left solely in the hands of individual parents and their doctors.

Adult screening, amniocentesis, and prenatal surgery are not the only areas where genetic engineering and selective breeding are starting to emerge as an unofficially accepted policy. In Chapter III AID sperm banks were discussed. One of the earliest supporters of AID, Nobel Prize winner H. J. Muller, advocated a program of identifying superior men in society, whose sperm could be frozen and stored in volume in sperm banks and then be used for impregnation as a way of improving the genetic pool of the population. Muller's list of the types of individuals whose sperm should be used has been a subject of controversy ever since Lenin appeared in his first list and was subsequently removed during the Cold War.[7] Nonetheless, his concept of "superior sperm" has been gathering steady support over the past several years. One such supporter is Sir Julian Huxley, who asks why AID sperm donors should be limited only to anonymous medical-school students, as is the current practice. Huxley believes:

> There should be a register of certified donors kept by medical men which would give particulars of their family histories. This would enable acceptors to exert a degree of conscious selection in choosing the father of the child they desire and so pave the way for the suppression of blind and secrecy-ridden AID by an open-eyed and proudly accepted EID where the "E" stands for eugenic.[8]

With AID already an accepted practice, what parent, confronted with a choice of being impregnated with an anonymous sperm or the sperm of a renowned artist, scholar, entertainer, athlete, or public leader, wouldn't prefer to select from the latter list? In this consumer-oriented society, the availability of these kinds of choices represents the ultimate consumer trip.

The only real obstacle to "celebrity seeds" is likely to be deciding whose sperm should be made available. Some say that a "national pedigree board would be required to decide who were to be donors."[9] Others favor such decisions being left to the marketplace function, with private sperm-bank corporations buying various donor sperm for whatever price the celebrities can command. The problem with this approach, say the "idealists," is that such costs will then be passed on to the consumer in the form of a sliding price scale for various types of celebrity sperm. This, they contend, will penalize prospective parents in the lower income brackets.

All of this is just the beginning, according to the new school of genetic engineers. The fact that there are already eight different ways to make babies (some already accomplished and the others in immediate sight) dramatically challenges the very concept of parenthood and family as we have known it through the ages. Bioparents, professional parents, single parents with cloned offspring are all new ideas on the horizon, and some scientists suggest that, for a variety of reasons, many people would favor these new forms of childbirth under certain circumstances. For example, if faced with the prospect of losing her job because of pregnancy, would a prospective mother choose to have her own fertilized egg removed and grown artificially in a laboratory? With 40 percent of the work force composed of women, many of whom are single parents already supporting a family, the temptation (or necessity) is compelling. For the wealthy jetsetter who would rather not interrupt her busy social schedule with nine months of pregnancy, the prospect of hiring a surrogate mother for her fertilized egg is not altogether unthinkable—at least, no more unthinkable than a wet nurse would have been seventy years ago. For some of the hundreds of thousands of homosexual Americans who would like to have their own child, cloning might well appeal as a viable option.

Dr. E. S. E. Hafez, experimental biologist of Washington State University, has even suggested that in the near future a housewife might walk into a special family medical clinic and browse through the aisles, looking at row after row of packets containing frozen embryos, each listing the probable characteristics of the child. She would purchase the embryo of her choice, pay for it, and then make an appointment to have it implanted in herself or in a surrogate mother or in vitro.[10]

The final word on the subject of childbirth and

genetic engineering, however, must be reserved for Dr. James Bonner of the California Institute of Technology. In an article published by the National Aeronautics and Space Administration Dr. Bonner asserts:

> The logical outcome of activities in modifying the genetic make-up of man is to reach the stage where couples will want their children to have the best possible genes. Sexual procreation will be virtually ended. One suggestion has been to remove genetic material from each individual immediately after birth and then promptly sterilize that individual. During the individual's lifetime, records will be kept of accomplishments and characteristics. After the individual's death, a committee decides if the accomplishments are worthy of procreation into other individuals. If so, genetic material would be removed from the depository and stimulated to clone a new individual. If the committee decides the genetic material is unworthy of procreation it is destroyed. . . . The question is indeed not a moral one but a temporal one—when do we start.[11]

CHILDHOOD EDUCATION

French biologist Jean Rostand talks about the possibility of literally doubling the size of the human brain to create a race of superior minds.[12] Noting that a large brain might pose problems in mobility, Dr. James Bonner suggests that a way be devised so that such abnormally large brains could "stay comfortably in one place and send their sense organs out into the world."[13] Parents who might not be ready to accept such possibilities—even though they are no longer just science-fiction scenarios—are much more likely to support new genetic injections that can be administered to their children to help improve their memory, attention span, and personality. These kinds of genetic injections are being looked into by scientists and being talked about by educators as a way of improving the quality of our childhood education and the general intelligence of our population. Dr. David Krech, professor of psychology at the University of California at Berkeley, says that "in the not too distant future they may well be talking about enzyme-assisted instructors, protein memory consolidators and antibiotic memory repellers" in education programs.[14]

Jargon aside, what parents wouldn't want their child to "fit in" or perform better—especially if such injections were being administered to other kids who were outperforming their own? In fact, proponents of gene therapy argue that children already rely on calculators, computers, video tapes, and an assortment of other "external" aids to enhance their learning abilities. Gene therapy, they contend, is merely an "internal" aid designed to help achieve the same end.

A solid groundwork for the introduction of mind-changing genetic technology has already been laid in American schools. Today upward of one million schoolchildren are being given special drugs to alter their behavior in the classroom. The two most popular drugs administered are Ritalin, manufactured by Ciba Geigy Corporation, and Dexedrine, produced by Smith, Klein and French. Ostensibly the children who are administered the drugs are diagnosed as exhibiting minimal brain dysfunction (MBD), which covers everything from a short attention span to difficulty in learning or responding to discipline. Even though the FDA has finally outlawed the use of the term MBD in advertising these drugs, because its meaning is too vague and because there is *no* evidence of any organic disease, schools continue to prescribe such medication at an alarming rate. Some authorities say that the number of diagnoses and subsequent medical prescriptions is doubling every two to three years. In many school districts 10 to 15 percent of the students take these drugs as part of the school program. For example, 10 percent of Oklahoma's sixty-two thousand public school children were taking Ritalin for MBD in one recent year, and in Massachusetts studies suggested that 25 percent of the youngsters are victims of minimal brain dysfunction.[15]

The main difference between these current drugs and future genetic injections, as Jon Beckwith of Harvard Medical School points out, is that with the latter, the process of altering the mind and behavior is irreversible. Once you've changed the gene, there's no going back; it's permanent.

In other words, with gene therapy, rigid biological limitations are artificially imposed on the schoolchild, limitations that will affect every aspect of later social and economic life in society. Critics like Beckwith are concerned that genetic injections of schoolchildren will be applied selectively to create a kind of biological tracking system in the schools, whereby certain children are genetically upgraded to prepare them for elite positions in society while others are downgraded to prepare them for menial job roles in the economy.

On the other hand, since most urban public high-school systems are already divided into vocational and college-oriented tracks anyway, wouldn't educators, and even parents and their children, conceivably welcome certain "biological enhancers," especially if they thought it would improve the individual's chances of getting ahead later in life? For example, would a student preparing to be a highly skilled mechanic turn down the chance to receive a special gene injection to improve potential manual dexterity? Or would a budding young English major reject the opportunity of being given a gene infusion to upgrade potential writing ability?

Wouldn't both these students think twice about the consequences of not having these gene implants listed on their future employment résumés—regardless of their effectiveness—knowing that it might well make the difference between their being chosen over someone else in a later job interview?

Critics also argue that gene injections might be used as a means of enforcing a biological social control over the school population, especially in inner-city ghetto schools and other so-called problem areas. Still, faced with escalating crime, delinquency and other forms of antisocial behavior in the schools, how many educators would resist the temptation to apply gene injections, if available, to arrest aggressive, hostile behavior? Especially since they are already using drugs and school security police to achieve the same ends.

Tracking and control arguments aside, proponents contend that in the final analysis the "cost factor" must ultimately be taken into consideration. The expense of maintaining existing educational services is rising dramatically throughout the country, while the overall quality of public-school education is declining. Beleaguered taxpayers, confronted with the prospect of either doling out more hard-earned dollars to upgrade educational services or accepting the introduction of cheap genetic technologies into the school system, will, say the proponents, be more than receptive to the latter proposition.

THE CONSUMER MARKET AND MEDICINE

Genetic engineering is likely to have a profound impact on every major consumer market. Take, for example, the deodorizing industry. With polluting smells becoming increasingly intolerable, some geneticists say that relief may be as close as a one-time gene injection to alter sensory perception. The Monsanto Chemical Company has already moved one primitive step in that direction with the introduction of a revolutionary new type of deodorant, which it plans to market soon. It's called Veilex, and instead of covering up bad odors with perfumed scents, it anesthetizes the nerves in the nose which sense bad smells, while leaving the good-smell nerves untouched.[16] Experimentation in altering the human response to smell is a relatively new field, but already the first taste-and-smell clinic has been set up at the Georgetown University Hospital in Washington, D.C., to conduct basic research.

Food is another market where genetic engineering is making inroads. Scientists are already busy at work trying to develop new types of grain that can be grown cheaper by taking nitrogen in from the air rather than relying on more expensive fertilizers. Meanwhile, scientists like Dr. A. M. Chakrabarty of General Electric are looking into the possibility of partially converting the human digestive tract (through genetic means) so that people will be able to eat cellulose (hay and grass). Of course, there is the possibility that such a procedure might result in the dangerous buildup of methane gas in the body, according to Chakrabarty. Nonetheless, "we think it is technically possible,"[17] he says, and claims that it could be done relatively simply by developing a special cellulolytic tablet which people would take every morning before breakfast. It might even be introduced, initially, as a new food alternative for welfare recipients on food stamps. With the rising cost of food of concern to most American families, Chakrabarty and others believe that eventually many people would be receptive to the cheaper hay- and grass-derived products, once the initial "stigma" had been overcome by appropriate promotion, marketing, and packaging techniques.

Genetic alteration is also expected to be enthusiastically embraced by millions of Americans who are concerned about improving their physical appearance. In 1976 alone, Americans spent over six billion dollars on some twenty-five thousand cosmetic products in order to improve their appearance.[18] In addition, over one million Americans last year underwent the more expensive "image-enhancing" operations.[19] Some researchers believe that cosmetics and plastic surgery will soon be looked back on as rather crude art forms when the science of genetic modification enables the American consumer to change such things as hair color, skin pigmentation, height, and general body structure. In the words of the Nobel Laureate Alexis Carrel, we are about to become "both the marble and the sculptor."[20]

Many scientists even claim they will very shortly be able to slow down old age and extend life way beyond our wildest imagination. According to Dr. Kendall Pye, a leading scientist in the field of enzyme technology and co-director of the Enzyme Technology Group at the University of Pennsylvania Medical School, researchers now think that aging might be linked to damage done to DNA molecules, occurring slowly over a person's lifetime. This damage, they believe, may prevent the cell from accurately reading the genetic information, which results in the production of relatively inactive enzymes that cannot function effectively. By repairing the damaged DNA molecules "some experts believe that it might then be possible to extend the average life span to well over 100 years—possibly several hundred years."[21] As some scientists have observed, it's almost impossible to imagine people rejecting such genetic technology once it becomes available to them.

Another area where genetic engineering is apt to find a ready market is so-called mood alteration. Says one scientist, "those of us who work in this field see a developing potential for nearly total control of human emotional status, mental functioning, and will to act."[22]

As geneticists learn more about the workings of the brain and the effects of polygenic traits on human behavior, they will be able to synthesize particular mood patterns directly into the biological structure of human beings.

The consumer market for mood-altering genetic surgery is already so firmly established that even critics are pessimistic about the possibility of forestalling its widespread and immediate use as soon as it becomes available. Their pessimism is based on the fact that so-called psychoactive drugs (the precursors of genetic injections) have, in just a few years, become the most heavily prescribed drugs in the country. These psychoactive drugs, according to a study prepared under the auspices of the National Institute of Mental Health, are being taken by over 40 percent of the women and 22 percent of the men in the U.S. today.[23]

While many studies have questioned whether some of these mood- or behavior-changing drugs have any clinical effect at all on the users, the important thing is the rationale that underlies their promotion, sale, and consumption. For example, Ciba Geigy says that its antidepressant, Tofranil, will help you readjust when "losing a job to the computer may mean frustration, guilt and loss of esteem."[24] Merck's Triavil is for people who are "sad or unhappy about the future . . . easily tired and who have difficulty in making decisions."[25] Not to be outdone, Serentil, by the Sandoz Corp., claims to be: "For the anxiety that comes from not fitting in. The newcomer in town who can't make friends. The organization man who can't adjust to altered status within the company. The woman who can't get along with her daughter-in-law. The executive who can't accept retirement."[26] According to the Sandoz Corporation, "these common adjustment problems . . . are frequently intolerable for the disordered personality."[27] Serentil, says the company, benefits such disordered personalities.

In a society where millions of people feel increasingly alienated at the hands of the giant bureaucracies that regulate their lives and feel more and more powerless against the haunting specter of pollution, urban decay, unemployment, inflation, and a host of other critical problems, psycho-drugs offer the ideal fix. Their promotional intent is to convince the American public that their problems are not social, economic, and political, but are personal and, more important, *biological.*

Nowhere is this market psychology of personal "biological unfitness" exploited more effectively than with women. Sixty-seven percent of all psychoactive drugs are sold to women.[28] For example, Hoffman-LaRoche claims that its drug Valium is for the housewife who has "psychic tension and depressive symptomatology . . ."[29] Another LaRoche drug, Librium, is for the college woman when "her newly stimulated intellectual curiosity may make her more sensitive and apprehensive about unstable national conditions."[30] The point made in all of these ads is that women's mental and emotional states stem more from their own biological disorders than from the values and institutional relationships that affect them in the outside world.

The incredible success of psycho-drug promotion and marketing has already paved the way for the introduction of mood-altering genetic engineering on a mass scale. Any consumer who has become habitually addicted to the purchase and consumption of psycho-drugs is a ready market for a "genetic fix."

Mass genetic injections and surgery are also likely to be heavily promoted by the medical fraternity. As was already pointed out, each of us carries a genetic load of somewhere between three and eight defective recessive genes, and there are more than two thousand different human disorders which are the consequences of genetic defects.[31] On the surface, these kinds of statistics present a compelling argument for mass genetic surgery, according to the enthusiasts. Of course, since everyone has some bad genes, it would become necessary to pick and choose surgical operations on the basis of some scale of priorities. If past history is any guide at all, however, there is a very real possibility that such surgery will become a cure-all for whatever ails you—especially when scientists begin to perform surgical procedures for "polygenic abnormalities" which involve moods and behavior. Today almost one half of all visits to the doctor's office are motivated by psychological and not physiological problems.[32] At present, doctors often prescribe psycho-drugs for these emotional problems, but it's not too difficult to imagine the same doctors recommending the more expensive gene surgery once it's available. If there are those who doubt that doctors would perform such surgery, especially if unwarranted, they might be interested in a Congressional report that found that in 1974 alone, doctors performed 2.4 million unnecessary operations, resulting in 11,900 unnecessary deaths, at an unnecessary cost to the public of $4 billion.[33]

LABOR

The National Institute for Occupational Safety and Health recently issued a report that found that over six hundred thousand American workers are being exposed every day to chemicals and other substances that cause cancer. According to the agency, another fourteen million workers are daily subjected to other toxic environmental hazards.[34] The costs of eliminating the hazards would run well into the tens of billions of dollars, and thus far industry has shown little inclination to bear the expense. Instead, many corporations are beginning to look seriously into the possibility of adapting the workers to the existing pollution. The theory underlying this "adaptation" model is that certain workers, because of their genotype, are more susceptible than others to the harmful effects of particular pollutants. Therefore, the implementation of industrial screening programs could go a long way toward matching the best genotypes to each job task. Dr. Charles R. Shaw of the University of Texas says that "the matching of genotype to job will someday be done routinely,"[35] and several companies have already established their own testing programs. One of these companies is Dow Chemical. The Texas division of the firm began industrial screening procedures over twelve years ago. The Dow medical team says that "persons under consideration for employment are given a pre-employment physical examination that includes chromosomal analysis . . . the employees are evaluated for cytogenic anomalies prior to work assignments."[36] The next step in Dow's industrial screening operation is to streamline the entire screening operation by developing "programs for completely computer-controlled analysis."[37] While there might be some initial resistance to such programs, employers are confident that workers faced with the prospect of contracting a serious illness or losing their job altogether will decide instead to accept genotyping procedures.

Many personnel executives are now looking toward the day when genetic therapy or surgery will be used to literally retailor the individual worker's biological make-up to fit it precisely into the total job role. Having workers who "fit" has been an issue of increasing concern to industry in light of the steady decline in worker productivity over the past decade. The problem of labor productivity has become so acute of late that business leaders are warning that if a way isn't found soon to reverse the trend, American workers will be increasingly replaced by foreign labor over the next quarter century. In fact, this is already happening on a large scale. Today one out of every three workers of an American-based multinational corporation is a citizen of another country.[38] Strangely enough, the American worker is the first to admit that on-the-job performance is poor. According to one Gallup Poll, 57 percent of the public thought "they could produce more each day if they tried," and a Harvard Business School Study found that industry frequently uses only 30 to 40 percent of the average worker's mental and physical ability.[39]

Probing into the reasons for this problem, an exhaustive report done by HEW found that "the discontentment of trapped, dehumanized workers is creating low productivity."[40] The increasing alienation of the American work force, says the HEW study, is making itself felt in a number of ways, from increased absenteeism to sabotage, alcoholism, and a general decline in mental health. The effect on productivity is staggering.

While some corporations have attempted to address the problem by implementing various piecemeal worker-participation and job-enrichment schemes, most have chosen instead to begin experimenting with different types of psychological and physiological conditioning programs to "subconsciously" manipulate the workers into improving their output. In fact, a new professional group, composed of ergomonists, anthropometricians, sensory-environmentalists, biotechnicians, and industrial psychologists, has emerged over the past several years, its job to design and implement such programs.

The conceptual thinking behind these conditioning efforts is already so closely related to the notion of gene therapy that it is setting a precedent for broad-scale bioengineering at the workplace as soon as it becomes feasible. Personality testing and psychological screening and counseling have already become an integrated part of most industrial personnel programs. Other major firms, such as Emery Air Freight, Questar, and Sohio, have introduced sophisticated "behavior modification" programs into the workplace.[41] Many companies are relying increasingly on such things as noise, lighting, and colors to evoke various physiological and mental responses from their employees. Some firms have introduced certain kinds of background sounds (called masking) to upgrade workers' physiological performance. The most effective is called "white sound," which is similar to the sound of a water fountain. Music is also used to stimulate productivity. The Muzak Corporation claims, for example, that it can use various combinations of music and rhythm to "change your heartbeat, . . . affect your metabolism and respiration . . . increase or decrease energy and alertness and make you feel relaxed or excited."[42] Studies have shown that such techniques increase productivity up to 15 percent or more.

Lighting is also being used as a conditioning agent, especially ultraviolet rays. Researchers have found that when ultraviolet rays hit the photoreceptors in the eye, the result is the secretion of the hormone melatonin, which is vital to health and sexual vitality. In tests using ultraviolet and full-spectrum lighting, subjects exhibit a greater attention span. Says John Ott, director of the Environmental Health and Light Research Institute, "We have finally learned that light is a nutrient much like food,"[43] and consequently industry is now making it a regular part of their employees' diet.

Even colors are being used to stimulate certain physiological responses. Specialists in this field have found that orange-reds most consistently maintain alertness, that cooler colors make the environment less distracting, that green is most conducive to meditative work, and that solid reds incite activity and emotionality.[44]

Since there has been very little resistance to the introduction of these kinds of techniques and programs, some personnel directors and efficiency experts confidently look to the time (in the near future) when gene therapy can similarly be introduced into the workplace as a tool to upgrade worker performances. While polygenic gene infusions could be invaluable in altering certain attitudes, such as hostility and anti-authority feelings, frustration induced by boredom, alienation, and job stress, they might also become a sheer necessity, argue the advocates, to upgrade the mental ability of the work force to keep abreast of the lightning advances in technology that are expected between now and the year 2000. As it stands, says James Mitchell, executive director of the Public Personnel Association and former associate director of the National Science Foundation, "we may eventually spend 40 percent of our work time training to keep up with the changes brought about by the rapid advances of science and technology." Mitchell believes that "the traditional educational system with [its] rigid instructional pattern . . . cannot cope with changes that occur so rapidly as they do in today's society."[45] Since, as Mitchell and others contend, organizations frequently cannot wait for an outside training program to be completed in order to begin a project, chances are that when the technology exists to synthesize things like improved memory retention directly into the brain cells at a fraction of the time and expense involved in traditional retraining programs, they will be taken advantage of, especially if it means gaining a competitive advantage over other firms and a wider profit margin as a result of improved efficiency and productivity.

Redesigning the workers is only one of the possibilities for the future. Buoyed by the success in teaching lower animal species like dolphins to do menial job tasks, many scientists say that it is now quite possible to genetically upgrade animals to enable them to perform many of the jobs people do now. Bruce Wallace, professor of genetics at Cornell University, says that the technology for breeding animals has "simultaneously made gorilla/man and chimpanzee/man hybrids possible."[46] Joseph Fletcher talks of producing chimeras or parahumans to do dangerous or demeaning jobs, and Wiegel and Tinkler suggest that special clonal beings—identical twins with specially designed genotypes—could be produced in large volumes to do similar tasks.[47] Middle-class taxpayers are also likely to show some interest in substituting chimeras, clones, or parahumans for some public-employee jobs, such as garbage collector, if it means a reduction in their tax burden.

THE MILITARY AND GENETICS

Nowhere is there more interest in the possible uses of genetic knowledge than at the Department of Defense. Writing in *BioScience,* Gairdner B. Moment asks rhetorically, "[can] anyone doubt that if a desperate major war should erupt, microbiological warfare based on these [genetic discoveries] would not occur?"[48]

The possibilities of genetic warfare are limited only by the imagination of the generals. Nobel Prize–winning biologist Salvador Luria says that in the future "we may witness efforts to invent viruses that can spread in an enemy population, genes that produce sensitivity to poisons, or susceptibility to tumors, or even transmissible genetic defects—in other words, genetic genocide."[49]

In a recent issue of the *Military Review,* published by the U.S. Army Command and General Staff College, a piece entitled "Ethnic Weapons" explained in detail the opportunities that new genetic knowledge would open up for military applications. According to the Army: "Observed variations in drug response have pointed to the possibility of great innate differences in vulnerability to chemical agents between different populations . . . It is quite possible to use incapacitating agents over the entire range of offensive operations from covert activities to mass destruction."[50] So far as the Department of Defense is concerned, genetic-warfare experiments are merely an extension of its long-standing testing program with biological and chemical agents.

For example, during the 1950s and 60s, the Army conducted secret experiments with LSD on over fifteen hundred unsuspecting soldiers and civilians in this country. The Air Force and CIA conducted similar covert experiments during the same period.[51] In December, 1976, it was also discovered (by *Newsday,* and subsequently confirmed by the Department of Defense) that between 1950 and 1966 the Army sprayed eight American cities and military installations with biological agents, to test the effects of possible future biological-warfare campaigns on densely populated civilian areas. In one joint operation with the CIA, the Army sprayed the New York subway system with bacteria "to provide a means of assessing the threat of infection to subway passengers." Although the military denied that the covert spraying was dangerous, authorities in one of the cities, San Francisco, said that it had resulted in an outbreak of infections, causing some illness and at least one death.[52]

Genetic technology has other possible military applications as well. For example, rather than risk the tragic loss of tens of thousands of young American men, the Army might well decide to produce entire divisions of chimeras, clones, and parahumans to fight in the next Vietnam-style war.

Reminiscent of the military's earlier campaigns around the space gap and the missile gap, a new term, the "gene gap," is just beginning to appear in specialized journals and in discussions among military strategists. The "gene gap," say these reports, poses a potential danger to our national security every bit as critical as the gap in nuclear preparedness. "Sooner or later," argues geneticist N. J. Berrill, "one human society or another will launch on this venture, whether the rest of mankind approves or not. If this happens, and a superior race emerges with greater intelligence and longer lives, how will these people look upon those who are lagging behind?" Berrill concludes with a dire warning that "they, not we, will be the heirs to the future and they will assume control."[53]

Of all the possible scenarios presently under discussion by advocates of genetic engineering, the "gene gap" could well be the one that is most likely to mobilize the entire society, virtually overnight, into the bioengineered Genetic Age. The launching of Sputnik and the effect it had on marshaling the resources, talent, and energy of America in an effort to overtake the Soviets in outer space should be compelling enough evidence of the potency and appeal of the "gene gap" argument.

Even without the gene gap, the military types have assembled an arsenal of emotionally laden arguments in support of genetic engineering that are apt to find a ready audience. For example, what would be the response of the public if a nuclear accident or limited nuclear exchange were to expose millions of Americans to deadly radioactive fallout? The effect on the present generation and on the future gene pool would be cataclysmic. Surely any national emergency imposed by the federal government would include a crash scientific program to find ways of mitigating the genetic damage by altering genes so that present and future generations could withstand radioactivity. In the event of an all-out holocaust, some contingency plans also include the use of new biogenetic technology. Sir Julian Huxley proposes that we "build deep shelters for sperm banks" and stock them with "collections of deep-frozen sperm (and ova) from a representative sample of healthy and intelligent" human beings. Huxley points out that "shelters for sperm banks will give better genetic results than shelters for people, as well as being very much cheaper."[54]

Unhealthy babies, poorly educated children, dangerous work environments, overpopulated cities, an increase in mental illness, and a weakened military are just some of the monumental problems facing this society over the years ahead. Therefore, the question becomes this: Assuming that people will be less and less willing to tolerate the rapid decline in their quality of life, which course of action is likely to have more success in gaining the approval both of our governing economic and political leaders and the general public: the basic restructuring of the institutions themselves, which would necessitate not only a radical shift in the control and distribution of power but also the spending of hundreds of billions of dollars to improve workplace conditions, health care, education, city life, et al.; or "adaptive" gene therapy to allow millions of individual Americans to adjust quickly and with the minimum of discomfort to the conditions that exist around them?

6
scientists and corporations

CETUS: THE IBM OF THE GENETICS FIELD

Attending the first National Academy of Sciences forum on recombinant DNA and genetic engineering, in March, 1977, were many of the major scientists doing research in the genetics field. Time and again during the three-day gathering, these scientists admonished critics for their "paranoia" concerning possible corporate exploitation of this emerging technology. At the same time, the proponents argued vociferously that their research was motivated purely by humanitarian considerations.

However, a confidential in-house memorandum prepared by a newly formed West Coast corporation tells an entirely different story. The corporation is called Cetus, and its board of directors and advisers—all of whom have a significant financial interest in the firm—reads like a "Who's Who" of the leading scientific researchers in the recombinant DNA and genetics field.[1] If there were ever any doubt about either the "real" motives of some of the top scientists involved in this research or of the incredible commercial profit that could be made off their work, this private report lays them to rest.

The Cetus memorandum says that because of recent scientific breakthroughs it has become possible, for the first time, to give real meaning to the term "genetic engineering." "The significance of this *power* cannot be exaggerated," according to the Cetus corporation. The document goes on to discuss the important breakthrough in transferring genes from one species to another and contends that it is "difficult to communicate the magnitude of such an accomplishment to a nonscientific audience." As a consequence, the Cetus report skips over the technical talk and gets right to the main point. "It is only a matter of time," says Cetus, "before [the corporations] begin to exploit these developments commercially. A new industry with untold potential is about to appear," and Cetus has a head start on capturing the field.

For starters, says Dr. Ronald Cape, the company's president, "In the next 30 years or so biology will replace chemistry in importance in this country."[2] Businessmen like Cape are already predicting that as we run out of oil resources in the next twenty-five years, new genetically engineered compounds—bacteria, molds, and fungi—will increasingly be used to supply the economy with alternatives to oil-derived chemicals for making fertilizers, plastics, wash-and-wear clothes, pesticides, dyes, paints, and tens of thousands of other products. In fact, they argue that it is only a matter of time before the right kind of new living microbes can be developed and substituted for just about any chemical presently being used for industrial purposes.

Expanding on the incredible commercial potential that a monopoly over the field of genetic engineering by Cetus would mean, the authors of the report also claim that in the bio-medical sciences "this concept [genetic engineering] is so truly revolutionary that by the year 2000 virtually all the major human diseases will regularly succumb to treatment" by genetic procedures. The Cetus people assert that "the pharmaceutical industry of 1975 will either have changed its entire product orientation by the turn of the century or new, more imaginative industry leaders, moving swiftly into the new field now, will have taken over a commanding leadership." Cetus already has a good idea of where that new commercial leadership will come from. "The mission of the Cetus Corporation is the exploitation of this accumulated body of knowledge." To accomplish this, "Cetus has assembled scientists of world renown in developing the newest frontiers of genetics and large scale sophisticated instrumentation." Who are these renowned scientists? They include Nobel Prize winner in bacterial genetics Dr. Joshua Lederberg; Dr. Stanley Cohen, associate professor of medicine at Stanford University Medical School, and the acknowledged leader in the new technology of gene manipulation; Arnold Demain, professor of applied microbiology at MIT; Nobel laureate Donald A. Glaser, professor of physics and molecular biology at the University of California at Berkeley, etc., etc.

The Cetus memorandum is neither shocking nor atypical, says biologist Jonathan King of MIT, who contends that the motivations of research scientists are often more self-serving than humanitarian: "They [the scientists] are into making a living. Take the guy who is a senior vice president for Kellogg's Corn Flakes and is trying to make it up to the president. Why does he want to do this? Because he likes corn flakes? No, because he has been brought up all his life to want to succeed and to have status, to have prestige and to make more money, or be more powerful . . . Is science any different? No!"[3]

The question of motives and the determination of what kind of research becomes the subject of inquiry was the topic of Jon Beckwith's address before the American Society for Microbiology a few years ago. There to accept the Eli Lilly Award for his work with genes, the Harvard biologist shocked the entire assemblage of dignitaries by contending that scientific inquiry and the scientists themselves are biased because "science [is] in the hands of the people who run our industries and is being used to exploit and oppress people all over the world and in this country."[4]

Ethan Signer of MIT's biology department agrees and adds that "the system" encourages scientists to conceptualize research in commercial terms.[5] These new advances in genetic engineering, say the critics, will not only "advance" the corporations that apply and exploit them, but also the individual scientists whose intellectual labor is used in the process.

These are heavy charges to level. Is scientific inquiry influenced and controlled by the economic interests that dominate American society? Are many individual scientists merely opportunists and an adjunct to the corporate payroll? Is recombinant DNA experimentation and research into genetic engineering being pursued in the laboratories mostly because of its enormous profit potential? How valid are the criticisms of King, Beckwith, Signer, and others?

FUNDING RESEARCH AND DEVELOPMENT

To assess how dependent the scientists are on big business one need look no further than the statistics on employment and on research-and-development funding. First, most scientists are directly employed in industry. In fact, in the next ten years over 90 percent of all college and university science graduates will work for the private sector.[6] Second, industry finances close to 70 percent of all the research and development in the country—which totaled a whopping $22.4 billion in 1974, the last year for which statistics are available.[7] Thus, the overwhelming majority of research scientists must depend directly on large corporations for their paychecks and their research grants. This in turn means that the type of research undertaken depends on the priorities established by the commercial interests.

It is also interesting to note that while 63 percent of industry's research and development is financed with its own funds, nearly 37 percent of the money is given to it by the federal government.[8] To insure a steady flow of government funds to industrial R&D programs, America's corporate giants make sure that most prominent "academic" scientists on the all-important government advisory committees are already on the corporate payroll.

Charles Schwartz, professor of physics at the University of California at Berkeley, has published a detailed study of the linkages between academic scientists serving on government committees and America's major corporations.[9] He has found that of the 78 scientists who served on the exclusive Presidential Advisory Committee between 1957 and 1973, 16 were directly employed in private industry at the time of their appointment, and another 4 held joint appointments in industry and on university faculties. While an additional 55 members were classified as "pure academics," a closer investigation by Schwartz found that over two thirds of them were actually directors or consultants of giant corporations, receiving remunerations of five thousand to ten thousand dollars or more for their affiliation.

Some of the "scientists" who have served on the Presidential Advisory Committee and who have also served on the board of directors of America's giant multinational companies are John Bardeen, a director of Xerox; Lloyd Beckner, a director of Texas Instruments; Harvey Brooks, a director of Raytheon; Harold Brown, a director of IBM; Melvin Calvin, a director of Dow Chemical; Caryl P. Hawkins, a director of Dupont; Donald Horning, a director of Westinghouse; James Killian, a director of GM; Franklin A. Long, a director of Exxon; George P. Paks, a director of McDonnell

Douglas; H. P. Robertson, a director of Northrup; Jerome Wiesner, a director of Celanese.

The President's Advisory Committee is not the only government body with such a tight interlocking relationship between "academic" scientists and corporations. There is also the "prestigious" National Science Board, which sets policy for the National Science Foundation, the government's science agency, whose purpose is to represent and fund "pure" academic research, independent of the influence of industry. Again, Schwartz has found that of the 78 scientists who served on this committee from 1957 to 1973, 62 were officially listed as "academics" when in fact more than half were also directors or consultants of giant corporations.

At the twenty-fifth-birthday festivities of the National Science Foundation, Dr. Paul Chenea, vice-chairman of General Motors, candidly summed up the function that "academic" scientists and government science bodies perform:

> Science and industry have always been clearly entwined with respect to each other's role in society. We in industry have come to depend upon funding scientific research and training in a broad spectrum of disciplines in universities and other similar organizations with the sponsorship of the National Science Foundation. NSF-sponsored contributions to science have been used again and again and are greatly appreciated.[10]

THE NATIONAL ACADEMY OF SCIENCES

Perhaps the best example of how industry has come to dominate science in this country can be found in a close-up examination of the prestigious National Academy of Sciences. The NAS was set up by an act of Congress in 1868. According to its charter, "the Academy shall, whenever called upon by any department of the government, investigate, examine, and report upon any subject of science and art."[11] While the expenses incurred by the Academy in doing reports were to be paid for by the federal government, the institution itself was to remain independent and autonomous. A former NAS president characterized the Academy as "a Supreme Court of final advice" whose work is "wholly in the public interest, uninfluenced by any elements of personal, economic or political force."[12]

Membership in the NAS is so exclusive that one recent survey of scientists showed that they considered admission almost as important as winning a Nobel Prize. Only 1,077 scientists, or approximately 1 percent of the total number in the U.S., are members of this august body. Membership statistics also reveal that over 70 percent of the Academy fraternity earned degrees at one of ten elite universities. The median age of members is sixty-two, and there are only two blacks and sixteen women in the total membership.[13]

Despite its claims to independence and objectivity in the name of pure science, the NAS has been attacked in recent years by a growing number of critics who believe that it has become a mere rubber stamp for establishment institutions. Former Secretary of the Interior Stewart Udall charges that the Academy has become a "virtual puppet of the government in a fashion that leaves little room to serve as an independent critical voice."[14] Even its current president, Philip Handler, acknowledges that if the Academy is "helpful" to the various agencies of government, then there is a greater likelihood that the government will put out a "larger welcome mat" for the scientific community's needs.[15]

Still, according to critics, the Academy's interest in "serving" the federal government is secondary to its interest in serving the American business community. Phillip Boffey, in his authoritative book on the NAS, entitled *The Brain Bank of America,* catalogues story after story of NAS complicity with and direct relationships to the giant corporations of the country.

Take, for example, the National Association of Engineers, a branch of the NAS. An Academy study done in 1971 showed that 43 percent of NAE's members were high-level corporate managers and another 22 percent were high-level executives in other fields.[16] This blatant example of corporate influence led *Nature* magazine to comment that "if there existed an organization such as a national academy of successful business and government executives, its membership would probably not differ greatly from the NAE." *Nature* went on to ask why, with all of the thousands of American engineers to choose from, the NAE continued to "draw its new membership . . . almost exclusively from the top brass at DOD, Ford, GM, General Dynamics, Lockheed and so forth."[17] In 1974–75, the president of the NAE was a former Secretary of the Air Force, the vice-president was a senior V.P. at Westinghouse, and its treasurer was president of General Motors. Other committee members were presidents or vice-presidents of companies like Exxon, Xerox, and Bechtel.[18]

Many of the NAS units receive much of their financial support directly from industry, which gives the business community the leverage it needs to influence the findings of reports and to determine what kind of studies are undertaken in the first place. This symbiotic relationship is particularly evident in the Drug Research Board, the Agricultural Board, Computer Science, Food and Nutrition, and the Building Research Advisory Board.[19]

Former FDA Commissioner James L. Goddard claims that the Drug Research Board is part and parcel of "the drug establishment, a close-knit, self-perpetuating power structure, consisting of drug manufacturers, government agencies and select members of the medical profession."[20]

Less kind in describing the Building Research Advisory Board is George B. Kistiakowsky, himself a former vice-president of the NAS. Kistiakowsky says the Board is nothing more than "an industrial lobby par excellence operating within the Academy structure."[21]

As for the committee that is supposed to deal with "automobile safety," the Highway Research Board, it should suffice to point out that its chairman is a top engineer at the Ford Motor Company and its secretary is an executive at General Motors.[22]

Then there is the Food Protection Committee, which has become a target of jokes even within the Academy membership. The committee is financed largely by the food, chemical, and packaging industry and is "biased as heck for industry," according to Dr. Marvin Legator, the FDA's former chief for genetic toxicology.[23]

Given the incestuous ties between the corporate world and the scientists who compose the various committees of the National Academy of Sciences, it should come as no surprise that of the nine financial sponsors of the NAS's first forum on recombinant DNA and genetic engineering, six were the very drug companies that are tooling up for entry into the genetics market.[24]

THE DRUG INDUSTRY AND GENETIC ENGINEERING

The Cetus Corporation is only the tip of the iceberg in terms of corporate involvement in genetic engineering. Major drug companies are already gearing up to enter this "revolutionary" new market, and giant chemical firms are expected to follow suit within a matter of months. The race to exploit genetic engineering is on, and the monetary rewards at the finish line promise to be nothing less than spectacular. Says *Business Week*: "The ultimate impact on the chemical and pharmaceutical industries could be similar to that which followed when an understanding of solid-state physics was brought to electronics: Genetic engineering is, in effect, ready to graduate from the vacuum tube to the transistor."[25] Critics agree, and argue that commercial involvement will not only result in a profit bonanza for capitalist America but will also set the precedent for corporate control over human bioengineering as well—control which they say will translate into massive abuse and exploitation of millions of American consumers and workers. While the specter of a corporate Brave New World looms as a major concern of the critics, industry representatives contend that such fears are totally unfounded and cannot be supported on the basis of past record and performance.

Yet a case history of pharmaceutical companies alone—the first major industry to enter this field—should be enough to convince the public that the critics are at least justified in their concern.

Annual sales in the drug industry approached fourteen billion dollars last year. Out of this total figure it is estimated that nearly 13 percent went directly to profits and another 20 percent into advertising to increase sales. For the past ten years the drug industry has ranked as either the first or second most profitable industry in the U.S.[26] Part of the reason for this lies in the fact that nearly three fourths of all prescription drugs sold in the U.S. can be obtained only from a single manufacturer, which allows firms to jack up their prices for drugs people believe they must have for their personal health.[27] The notion of profiteering at the expense of people's illnesses has not gone down well with the American public, which ranked the drug companies as the number two corporate villain (the oil companies rank first) in a recent nationwide public opinion survey.[28]

Excessive profiteering is only one side of the drug-company equation. The other is safety and effectiveness of products. According to an exhaustive study published by a Senate subcommittee in 1962, of the four thousand drug products legally marketed in the country over the past twenty-four years, almost half had no scientifically proven value.[29] Even more startling is the fact that many of these ineffective products, which are produced by major pharmaceutical houses, are actually dangerous and have caused illness and death. In their book, *Pills, Profits, and Politics,* Milton Silverman, research pharmacologist, and Philip Lee, former assistant secretary of HEW, report that the adverse effects of drugs "kill more victims than does cancer of the breast." The problem has become so acute, say the authors, that adverse drug effects now "rank among the top 10 causes of hospitalization and are held accountable for as many as 50 million hospital patient days a year."[30]

For those already concerned over the prospect of major pharmaceutical companies becoming involved in genetic engineering, the countless examples of unethical and criminal wrongdoing by every single one of the firms already listed as being involved in this field are not reassuring. Morton Mintz and Jerry Cohen in their book *Power Inc.* have catalogued many of the case histories.

For example, Abbott Labs, one of the firms engaged in recombinant DNA experimentation, recently pleaded no contest to a sixty-count conspiracy indictment brought by the Justice Department. The indictment charged Abbott "with conspiring to sell and with selling contaminated intravenous fluids that were associated with 9 fatal and 150 non-fatal blood poisonings."[31]

Another firm tooling up for genetics experiments, Eli Lilly, was sued in 1977 (along with the University of Chicago) for $77.7 million in damages by Patsy Mink, Assistant Secretary of State, and several other women. The suit claims that Eli Lilly, in cooperation with the University of Chicago, administered the drug DES (diethylstilbestrol) to over 1,100 women in an experiment conducted without their knowledge. The drug was later linked with cancer in their offspring. The suit accuses Eli Lilly and the University of Chicago of "intentionally concealing from plaintiffs that they had been given DES and breaching their duty to inform plaintiffs and their children of the experiments and of the precautions their children should take to minimize the risk of contracting cancer or other abnormalities."[32] Even more disturbing, despite the fact that for the past six years the FDA had repeatedly warned about the danger of DES, Eli Lilly refused to take the product off the market.

The first major drug firm to announce plans to build a maximum-risk recombinant DNA laboratory is Hoffman-LaRoche. This is the same company whose plant in northern Italy created worldwide headlines in July, 1976, when it released a deadly gas that caused over seventy-eight million dollars' worth of damage to the environment and exposed nearly a thousand residents of the surrounding area to severe illness. The released poison was so dangerous that an investigative commission of the Italian government recommended that all buildings in the exposed area be torn down and destroyed, all vegetation burned, and a foot of topsoil be skimmed off. Despite the fact the LaRoche officials knew that the deadly poison causes extensive kidney, liver, and lung damage and a greater damage to fetuses than thalidomide, they purposely concealed news of the leak from the public for over a week, which was more than enough time for it to do its deadly business. Months after the incident scientists from Italy, Sweden, Switzerland, Britain, and the United States still had found no way to remedy the incredible damage. Said

John Maniello, of the U.S. Embassy in Rome, "it's like fixing a car by the side of the road . . . You try one thing and it doesn't work, so you try another."[33]

Much of the success of the drug industry can be attributed to the hold it exercises over the American Medical Association. A good case in point involves the Upjohn Company, another one of the firms entering the genetics market. Not so long ago, a massive multimillion-dollar study financed by the National Institute of Arthritis and Metabolic Diseases found that drugs like Orinase and Tolinase, manufactured by Upjohn, increased the cardiovascular death rate 250 percent. This was unwelcome news at Upjohn, which was selling these drugs to over a million and a half people.[34] In an effort to save this hundred-million-dollar drug market, the chief executive of the AMA, Dr. Sammons, sent a letter to over four hundred medical associations in which he asserted that "a considerable body of expert scientific opinion contradicts these findings." Dr. Sammons went on to say that individual patients should "not be influenced by press reports" and should continue whatever drug program they were already on.[35] Upjohn immediately asked for and received permission to reprint Sammons's letter for use by its sales force in the field, even though AMA's own legal counsel said that the organization's name could not be used for trade purposes or for purposes that might be construed as an endorsement of a particular product by the AMA.

The Upjohn case is typical of the relationship between the AMA and the drug industry, according to testimony given by Drs. John Adrian and Harry C. Shirky before the Senate Small Business Monopoly Subcommittee. Adrian and Shirky both chaired the Council of Drugs of the AMA. In 1972, the Council had drawn up a report that claimed that dozens of popularly prescribed drugs were "irrational" and should "not be recommended." The AMA quickly moved to abolish the Council, which had been operating for over sixty-six years. At the Congressional hearings the past chairmen of the Council charged the AMA with being "a captive of and beholden to the pharmaceutical industry."[36]

That charge was amplified by evidence subsequently submitted to Senator Kennedy's Subcommittee on Health. The data showed, among other things, that the drug industry had, in 1973 alone, distributed over two billion free samples to physicians as well as thirteen million gifts totaling fourteen million dollars.[37]

In addition to all this, the AMA was found to have invested nearly $10 million from its members' retirement fund in the drug companies. Finally, it was disclosed that over a three-year period the American Medical Association Political Action Committee, which is the lobbying arm of the AMA, secretly accepted $831,000 in political contributions from twenty-seven of the major pharmaceutical firms in the country.[38]

Drug-company abuses have reached such astronomical proportions that, in the fall of 1976, the FDA requested and received a Congressional appropriation of $16.4 million to add an additional 600 persons to its staff in order to monitor the research sponsored by pharmaceutical companies. In announcing the new appropriation, the director of the Bureau of Drugs of the FDA, Dr. J. Richard Crout, said in a prepared release: "We have learned that we cannot accept the test results submitted by drug sponsors on faith alone . . . Recent history has convinced us that we must look more closely at the validity and quality of the research done by industry."[39]

Despite a forty-year track record of abuse, exploitation, corruption, and gross negligence on the part of the pharmaceutical companies, Congress and the executive branch of government have never questioned for a moment their right to move into the field of genetic engineering. In fact, when it came time to draw up guidelines for private-sector involvement in this field, the

NIH and other government agencies charged with the task of setting policy on this matter made sure that pro-industry people would be well represented in the deliberations. For example, a random check of the backgrounds of the representatives of the government's industrial guidelines committee showed that seven individuals had previously been employed with major U.S. corporations. Two had served with major pharmaceutical firms involved in the recombinant DNA field. Oswald Ganby, the State Department representative, was previously employed as assistant director of international relations at Merck, Sharp and Dohme labs; Department of Transportation representative William D. Owens was at one time a director of a subsidiary of the Searle Corporation.[40]

The pro-industry bias of the federal government's committee on recombinant DNA and genetic engineering is exemplified in the all-important question of whether drug companies (and other firms) have a right to patent the processes for "new forms of life." Not a single member of that committee questioned the right of private corporations to assume a proprietary ownership over any "new life forms" that they commercially develop.[41] This unanimous agreement over the patent issue was accepted virtually without reservation in House and Senate hearings on industry involvement in genetic engineering in the late spring of 1977, thus establishing a precedent which could have serious implications five or ten years from now, when corporations will be able to market many of the ideas and technologies covered earlier in this book.

THE IMPLICATIONS OF CORPORATE CONTROL OVER GENETIC ENGINEERING

That the scientists are on the corporate payroll in one form or another should not really come as any great surprise. As Robert Nisbet, professor of Humanities at Columbia, put it:

> America is an intensely political and commercial civilization. If the man of knowledge remained for a long time more or less immune, such immunity could not reasonably have been expected to last forever. Starting out with the intention of doing good, he wound up doing very well. He cannot really be blamed for yielding to the temptation of marketplace and throne.[42]

Nor, for that matter, should the public be shocked to learn that giant drug companies often abuse and exploit the public in order to make a profit. After all, the first and primary responsibility of private corporations is to expand and make profits for their owners; if situations arise in which the public welfare conflicts with the corporate welfare, there is little question of which side a corporate manager will come down on.

What is important, however, is the profound effect that the new technology of genetic engineering could have on this society when applied by the dominant institutions in American life today, the giant multinational corporations.

Consider the fact that two hundred corporations already own two thirds of all the manufacturing assets in the country and employ over twelve million workers. Twenty-nine corporations own 21 percent of all the cropland in America. Eight oil companies control 64 percent of all proven oil reserves, 60 percent of all natural gas, and 45 percent of all known uranium reserves. So powerful are these giant economic enterprises that of the one hundred largest "money powers" (measured by GNP or gross sales) in the world today, thirty-six are no longer even countries, but American multinational corporations. Through the control of production, capital investment, employment, market practices, and consumption habits, these colossal financial empires successfully manage the daily affairs of millions of people.[43]

As students of modern industrial society have pointed out, this awesome economic power is principally exercised through the use of new technologies. In the world of expanding economic bureaucracies, technology is the system used to develop "highly centralized and intensive control over large groups of men, machines and events by small groups of technically skilled men operating through organizational hierarchies."[44]

The objective of advanced technology is to minimize individual or group behavior which is erratic or otherwise not easily classified and to eliminate all forms of activity that are unresponsive to centralized management.

Up to now, technology has been used by industry both to redesign the external world and the work process and, through communications, advertising, and various types of rewards and punishments, to condition people's behavior in order to fit them into the demands of the system. Now, with genetic engineering, the economic institutions of our society have the technological power to redesign the internal world of human life.

Is there any reason to believe that the economic bureaucracies of our society will reject the "technological imperative" which has been the *modus operandi* and the driving force behind their expansion and success once it becomes applicable to redesigning and regimenting the most important part of the economic process, human beings?

As for the government's role in genetic engineering, it is likely to continue providing industry with R&D funds and will no doubt continue to fashion minimal regulations and controls over corporate research and reporting procedures. But it is not likely, in the foreseeable future, to take over the development and broad application of genetic technology.

In our capitalist system the task of applying new technologies has been traditionally reserved for the private sector. The only real exception to this is the military, which is likely to rely increasingly on the use of genetic technologies in its war-preparedness programs. Of course, it's not inconceivable that a President of the United States could become so enthralled with the possibility of genetic technologies as to begin to use bioengineering as a prime instrument of social policy and political control, with the support of other key sectors and interest groups. Short of that, genetic modification is likely to be relied on increasingly in the prisons, mental institutions, and old-age homes as a cheaper and more efficient approach in dealing with these major social problem areas.

However, broad-scale genetic engineering will probably be introduced to America much the same way as assembly lines, automobiles, vaccines, computers, and all the other technologies. As each new genetic advance becomes commercially practical, a new consumer need, either real or manufactured, will be exploited and a market for the technology will be created. Over a period of time (the next twenty-five to fifty years) the cumulative effect of this step-by-step process will be the emergence of a kind of corporate Brave New World, not unlike the one Aldous Huxley fantasized about over forty years ago. It is a much less dramatic approach to the ultimate enslavement of the human species, but the results are no less terrifying than if had they been ruthlessly imposed by some mad political dictator. The only real difference is that with this approach we will march passively and in some cases even willingly into this new reality without pain, without inconvenience, and without awareness.

7

at the crossroads of human history

Social evolution follows a kind of cyclical projectory, just as natural evolution does. Cultural realities are punctuated by birth, life, and death cycles. In the social world as in the physical, a process of continual editing and revision is constantly forcing prevailing realities to be absorbed into or replaced by newer ones in the ongoing process of social change. While the editing process is subtle, almost imperceptible, the cumulative effect of the many revisions and mutations, one heaped upon another, is always some kind of critical watershed where new realities meet old. At these junctions, the body politic finally becomes aware that the old order that it was nurtured on is dying; that the social and economic relationships that gave shape and meaning to its culture are atrophying while several new realities are emerging from deep within its bowels, each ready to assert its right to be the natural successor in the chain of social evolution. The question of which new reality finally survives depends on the competing visions and determination of the human actors involved. Our country faces just such a watershed now.

America's powerful industrial machine is slowly grinding down. Our rate of growth has plunged to one of the lowest levels in the Western world. West Germany, Italy, and Japan now have a per capita real growth rate twice that of the U.S. French growth has been two-thirds greater than ours and even Britain has grown 10 percent more than the U.S. in the past twenty-five years.[1] Our government tells us that within ten years the nation will run out of the basic fuels necessary to sustain America's industrial base at required operating levels. The energy crisis has reached such critical proportions that the President of the U.S. contends that only a wartime-type mobilization can save us from collapse. At the same time, scores of America's major cities teeter on the brink of insolvency as municipal governments frantically slash vital public services such as sanitation, public schools and libraries, fire and police protection in an effort to buy a few more years of operating time. One study just released by the League of Cities says that eight of the largest urban governments in the country—New York, Buffalo, Detroit, Newark, St. Louis, Boston, Cleveland, and Philadelphia—could collapse financially at any moment.[2] Meanwhile, American-based multinational corporations have begun to retire many of their domestic plants and facilities, while simultaneously transferring their capital and investment funds abroad, where they are blessed with easier tax burdens and cheaper labor. In 1960, U.S.-based global companies were investing 11.5 cents for new plants and equipment overseas for every dollar similarly invested at home. By 1975, they were placing 31.9 cents abroad for every dollar of new investment for plants and equipment in the domestic economy.[3] At the same time, control over the American economy continues to be concentrated in the hands of fewer and fewer giant corporations. Today, fewer than fifty private institutions control over eight hundred billion dollars, which is nearly twice as much money as the entire federal government controls for fiscal 1978. Meanwhile, the effect of a hundred years of unplanned industrial expansion has taken its toll on the environment. Pollution has become so bad that our rivers and lakes are dying, our plant and animal life is becoming more and more diseased, and our urban sprawls are subjected to week after week of pollution alerts as citizens are warned to stay inside behind closed windows and doors.

Accompanying this decline in the economic quality of life is a growing deterioration in the social life of the nation. In a culture that has become increasingly characterized by mass phenomena—mass bureaucracies, mass advertising, mass spectator sports, mass consumption habits—millions of Americans feel increasingly alienated from the workings of their society and have come to feel more a target to be exploited by the powers that be than an active participant in the social process. As a result, alcoholism now claims over nine million Americans; mental illness an additional two million per year. Over nine million individual acts of crime were committed last year in America, including nearly a million crimes of violence. Fear of our neighbors has turned America into an armed camp, with people buying seven million new guns each year.[4]

The sense of personal futility and despair has now reached the point where most Americans no longer even attempt to seek a redress of grievances through the traditional political channels. Nearly half of the eligible voters in the country no longer exercise their franchise. According to one survey done by Louis Harris, less than 7 percent of the public believe their opinion counts for anything in the government decision-making process. Another survey found that over 50 percent of the people who do not vote believe that more "radical" approaches to change will be necessary.

From this crisis, two schools of revolutionary thought are emerging. While both agree that fundamental changes will be required over the next twenty-five to fifty years, one claims that what must be changed is the biological make-up of the human species to adjust to the rapid changes going on in the external world, while the other argues that the institutions and values of society are what must be refashioned.

The new eugenicists argue that a good part of the reason for the crisis facing civilization can be found in our biological imperfection. "Human culture has grown so rapidly," says Joshua Lederberg, "that the biological evolution of the species during the last hundred generations has only begun to adjust to it."[5] "The only exit from this cycle," says another biologist, is "to change man himself to escape the tyranny of genetically patterned development."[6]

In contrast to the genetic engineers are the more traditional social activists, who contend that the problems facing civilization are of an institutional nature; that power and decision-making are concentrated in too few hands and the rewards of society are unequally distributed. The social activists argue that a change in values is needed to emphasize qualities like cooperation, conservation, and participation in fashioning new institutional and interpersonal relationships in society.

THE GENETIC ENGINEERS

Genetic engineers decry the unpredictability of the natural world. Social progress is to be measured by the conscious effort to eliminate from the environment whatever is perceived to be random behavior, erratic activity, or unnecessary error. The underlying assumption behind this perspective is that greater predictability means greater control. The greater the control, argue biological engineers, the greater the security. For genetic engineers security is the ultimate value upon which to rest a conceptual framework.

A genetic engineer is every bit an engineer in perspective and outlook, seeing the natural world as a giant factory of sophisticated biological machinery which, while working fairly well in an untouched state, is nonetheless not as efficient and streamlined as it could be if it were retooled to maximize performance.

Improving efficiency and productivity is what engineering is all about. Genetic engineers argue that the human race long ago made the decision that the world should be reengineered, and contend that civilization is the deliberate act of synthesizing, or "making over," the natural environment to serve the ends of social progress.

All engineers bring the same approach to their work—it's called technique. Technique is a process of bringing together measurable, concrete, quantitatively derived data in order to find the one most efficient way to do a certain thing. Technique deals only with the verifiable parts of reality. Since moral, ethical, and emotional phenomena cannot be measured in any concrete, objective sense, they play no role in the technological process. In fact, such considerations are a detriment to the technological process, for the very reason that they are unpredictable and subjective in nature. Nor can technique take into consideration future possibilities, since the realm of the unknown cannot be concretely measured or predicted. For these reasons, technique is a limited approach to dealing with the world, based as it is on assessing only part of what makes up the total reality.

Thus, when genetic engineers apply technique to engineering life, their concern is with rearranging cells, molecules, and chemicals in whatever fashion will best improve the efficiency of the organism. To genetic engineers, we are the sum total of our observable parts, nothing more, nothing less.

The genetic engineers argue that "man the machine" is not keeping pace with the advances that have been made in a larger environment that is becoming increasingly technologized. People are still imperfect, erratic at times, often unpredictable and prone to subjective miscalculations. This imperfection, they argue, has a dysfunctional effect on the rest of the techno-system. If we are to prevent the entire system that we have now synthesized from collapsing in on itself, as it is showing signs of doing, then the only hope, they say, is to bring the last major component of the system into line with technical design. That means us.

In the battle between those advocating a biological revolution and those supporting a social revolution, the genetic engineers enjoy a distinct advantage. First, they are supported by all of the immense power of the prevailing economic interests in America. Power over communications and advertising, power to influence the political machinery, and power over employment are all formidable tools for securing public acceptance of genetic engineering. Second, there is the "expediency" factor. Mobilizing for basic social change requires a long-range commitment and sacrifice by millions of people working together with a shared sense of common purpose. Changing values and restructuring institutions is a painful process. In the short run, people's daily lives are impinged upon, familiar habits are upset, and traditional ways of doing things are challenged. David Baltimore, one of the more vocal advocates of genetic engineering, observes that "human beings are very conservative about their personal habits and do not easily change them." Since "it is unlikely," says Baltimore, "that we are going to be able to design a civilization that will be acceptable to the population,"[7] the alternative is to reconstruct the genetic composition of people themselves. What Baltimore and his colleagues are implying is that our society is already so fragmented and individuals are already so atomized that a collective mobilization for social restructuring is not very likely. Moreover, while they don't come right out and say it publicly, the genetic engineers are aware that the American public has over the years become conditioned to accept the short-run "technological fix" as an alternative to having to confront the long-range problems that continue to build up.

Nowhere is this psychology of short-term palliatives

versus long-range solution more apparent than in the health-care field. Although the U.S. spends over $100 billion per year on health care, it ranks fifteenth among the nations of the world in infant mortality and life expectancy.[8] Part of the reason for this lies in the fact that our health-care system is structured for short-term treatment of illnesses rather than long-term prevention of them. Despite the fact that hundreds of reports and studies have concluded that the major causes of death and disease are preventable and related to occupation, environmental deterioration, nutrition, consumer products, and poverty, only limited government or private funds are put into programs designed to address the problem. For example, in cancer research, where environmental factors are linked to 80 to 90 percent of all illness, the National Cancer Institute devotes only 15 percent of its $815 million budget to developing a better understanding of environmental prevention; the other 85 percent goes to research for curing people who come down with the disease.[9]

This lopsided focus on treatment rather than prevention is directly related to profits. Since most illness can be prevented by proper long-range health programs, an effective, publicly financed effort in this area would dramatically reduce the number of individual cases requiring medical treatment, which would, in turn, mean less profit for the medical industry. A shift in focus toward long-term prevention programs would also necessitate a complete restructuring of the entire health industry. For these reasons the medical lobby—doctors, drug companies, hospitals, private insurance companies, et al.—makes sure that short-term treatment rather than long-term prevention remains the *modus operandi* of our health-care structure.

If a cross section of Americans were asked if they believed that more attention should be focused on long-range prevention rather than short-term cures, they might well agree. But the fact remains that until people become ill they do not, and probably will not, devote a great deal of personal time and energy to thinking about health practices. When they do become sick, they want a treatment now, and are even less concerned with long-range prevention programs because for them the damage has already been done.

Advocates of genetic engineering are seizing on this "expediency" psychology to gain popular support for their research. Even though, as Jonathan King points out, in the five major health-problem areas—environment, infection, old age, poor nutrition, and genes—"the genetic components of illness still rank far below the others,"[10] bioengineers are busy convincing the public that its only hope in the short run lies in genetic cures. Every day or two a story appears in the press heralding a new genetic breakthrough, always with the tag that the experiment has brought medical science one step closer to curing cancer, diabetes, heart disease, and a host of other illnesses. As genetic treatments for specific ailments become practical, very few people are likely to reject them, especially if it means the difference between prolonged illness, suffering, and death or a measure of restored health. At the same time, the more people become addicted to the short-term genetic fix, the less likely that long-range prevention programs will be undertaken at all.

In every area of struggle between the genetic engineers and the social activists, the genetic engineers have the "expediency factor" on their side.

For example, would a family faced with rising food bills be more likely to mobilize with its neighbors, over the long run, to challenge the giant food monopolies that are hiking up prices and reaping windfall profits, or take a tablet once a day so that they could immediately begin to digest cheaper hay-derived food products?

Would factory workers be more likely to risk their jobs by organizing to demand a retooling of their asbestos plant so that the toxic substances are brought under control—a process which could take years to accomplish—or subject themselves to a painless gene injection making them immediately immune to the adverse effects of the chemicals?

The genetic engineers are betting that in these and thousands of other instances in which people have to choose between a long protracted struggle to change institutions and a short-term decision to slightly alter their genetic make-up so as to adjust to the existing order, they will inevitably accept the latter proposition with very little reservation.

THE SOCIAL ACTIVISTS

The social activists believe that people are more than just the sum of their chemical parts. To them human beings exhibit certain nonmaterial qualities which are not empirically measurable. These qualities include the capacity to feel emotions like love, compassion, and empathy. Unlike the genetic engineers, they believe that we already have within us the power to achieve perfection. But their definition of perfection has little to do with efficiency. Perfection, for the social activists, is the attainment of harmony and enlightenment; feeling a sense of oneness with the world. The social activists, while not discounting the use of technique in ordering certain parts of reality, place it in a different perspective from the biological engineers. Technique is not an ideology in and of itself, but one means among many for advancing inherent potential.

The social activists put a premium on individual freedom. Uniqueness, spontaneity, and nonconformity are not seen as qualities that are detrimental to human progress, but as their most necessary ingredients. When social activists have failed, it is because in the process of fashioning social arrangements the individuals involved lost sight of their initial inspiration. Still, the inspiration itself has a life independent of the time, place, and circumstances of the particular human actors. For this reason, people and societies can fail, but, it is contended, the underlying conception of our inherent perfectibility and enlightenment is a universal truth which continues to manifest itself at every moment of our existence, whether we choose to acknowledge it or not. The social activist believes that our society is writhing in crisis precisely because those universal truths have been ignored, that inspiration lost. The social, political, and economic relationships which were supposed to free the human spirit and provide a path to enlightenment have become obstacles instead, say the social activists. Our biology is not the problem, they contend; our motivations, values, and institutional relationships are.

The best illustration of how genetic engineering is being used as an excuse to avoid the real nature of the problems facing society, say the social activists, is recombinant DNA research. When proponents like David Baltimore of Harvard argue that if we give up recombinant DNA experiments we will be forced to live with famine and cancer, the critics reply that such statements reflect a refusal to deal with the underlying economic and political basis of our problems, and they offer the following examples: Baltimore and his fellow recombinant enthusiasts argue that their work might lead to the development of new food strains that could solve the problem of world hunger. Ethan Signer of MIT disagrees. According to studies done by the World Bank, the UN, and other international bodies, says Signer, there is already more than enough grain to feed the world's hungry, but people continue to starve because of maldistribution, which is a political and economic problem.[11] Again, Baltimore argues that we need recombinant DNA research if we are to find a cure for cancer. George Wald responds that most of the cancer in this country is of environmental origin and that science should be used to find new ways to clean up the environment rather than tinkering with the genetic make-up of people.[12]

Genetic engineering, say the social activists, is a logical extension of the notion that life has no value other than the utilitarian. This concept has so permeated our social reality, they assert, that everything, including people, has been reduced to an object to be turned on and off, exploited, used up, and replaced.

It is no wonder, they say, that people are suffering. In a society increasingly organized around a strictly technological imperative, human values are universally ignored in the decision-making process. There is simply no room for them.

The social activists argue that to reverse this trend we must restore lost values and reconstruct institutions so that people are allowed to participate and share more fully and equally in the responsibilities and benefits of human progress. If we don't act, they warn, we might well end up committing the ultimate folly—turning ourselves in for a less human but more efficient model.

Despite the fact that the momentum seems to be riding with the genetic engineers, the social activists contend that underneath the surface some profound changes have been taking place over the past decade in the attitudes and perceptions of many Americans—changes they believe could well signal a public rejection of genetic engineering and a new movement for basic changes in the society.

First, public perception of the role of science and technology has radically changed. "A few years ago," says Phillip Handler, the president of the National Academy of Sciences, "people regarded science and technology as a huge cornucopia that was going to enrich everyone's life. Now many people feel that science and technology have done more harm than good."[13] The public has good reason to feel that way. Americans have become painfully aware of some of the adverse effects of new scientific and technological advances over the past twenty-five years. Military technology has brought us to the brink of nuclear holocaust; industrial technology has wrought irreversible damage to the natural environment; the technologizing of life in general has resulted in an increasingly impersonal and robotized culture. According to one opinion survey, over 57 percent of the public believe that science has made life change too fast, and nearly half now believe that science has become a disintegrating factor. Between 1966 and 1973, public confidence in the people running the scientific community dropped 19 percentage points, to a low of 37 percent.[14]

The American public is also becoming increasingly hostile to the economic system that controls the development and application of most of the scientific and technological advances. The list of grievances has become almost endless: unsafe products permanently disable and kill tens of thousands of Americans each year; oil companies reap windfall profits off the energy crisis; corporations bribe candidates and illegally buy elections; chemical companies pour deadly substances into rivers and lakes; institutional pension funds siphon off the hard-earned retirement funds of workers, etc.

The abuses, the exploitation, the corruption, and the callous disregard for workers and consumers have taken their toll. The American people have not only begun to lose faith in the giant corporate institutions, but have also come to look critically on some of the essential features of the economic system. The concept of unlimited exploitation of resources, profit maximization, and monopolization of markets is being challenged as never before by an angry public. Negative feelings toward industry run so deep that, according to a Harvard Business School report, an overwhelming majority of top corporate leaders now fear that the public might demand a fundamental restructuring of the entire system within the next ten years.

There are also indications that for the first time the American public might be on the verge of challenging the orthodox view of progress. For example, a 1977 Harris poll reports that by a margin of 44 percent the public would choose "breaking up big things and getting back to more humanized living" over "developing bigger and more efficient ways of doing things." The same survey found that 74 percent of the American people now opt for "learning to get pleasure out of nonmaterial experiences" rather than "satisfying the needs for more goods and services."

The search for new modes of self-expression and new ways of looking at reality has led millions of Americans to take up a variety of non-Western experiences over the past five years, including meditation, yoga, and Zen Buddhism. Traditional organized religions have also shown an increase in active membership recently.

The growth in spiritual awareness has been accompanied by a growing interest in conserving the natural environment. The traditional notion of ruthlessly exploiting and controlling nature in the name of progress is being challenged by an environmentalist creed that emphasizes a reintegration with the ecosystem.

Distrust of unbridled scientific and technological progress, opposition to corporate hegemony over the life of the nation, a new spiritual awakening and concern for the natural environment are, say the social activists, creating the necessary conditions for a fundamental change in the values and institutional relationships of American society.

THE FUTURE

Whether the biological revolution or the social revolution prevails depends ultimately on how the American public responds. Choosing to restructure the institutions of society will require an affirmative act on the part of the public. On the other hand, genetic engineering, to succeed, will require only a passive acceptance.

No doubt some will harbor the notion that both realities can emerge and coexist with each other even if uncomfortably; or that it is not really necessary to entertain either vision at all. On this latter count it should suffice to point out that when those in power and those without power both begin saying that we must start to prepare ourselves for some profound changes in our way of life—even if they have different things in mind—our existing reality is in fact about to give way.

As for the argument that biological and social revolution can coexist, nothing could be more impossible. The assumption underlying social activism is that biologically we are sacred and inviolate. The assumption underlying genetic engineering is exactly the opposite. If genetic engineering were allowed to continue at its current pace, the Homo sapiens species would experience no more than five or six more generations before being irreversibly replaced by a new, artificially engineered organism. Though this new species would include some of our characteristics, it would in many ways be as different from us as we are from our closest relatives, the primates.

Some will argue that genetic technology, like other technologies, is in fact neutral and that whether it is used for good or ill depends on who is controlling it and to what end. In this society genetic engineering is being encouraged by the prevailing economic interests, who see in it a means of producing greater profits and increased technological control over society. Yet bioengineering can just as easily serve any institutional elite concerned with extending its power over human populations. For this reason, it matters little whether this technology is used in capitalist or socialist societies. While its application may differ from country to country, the end result will not. That is because inherent to the process itself is the concept of the engineers and the engineered. "It seems quite certain," says one observer, "that the new power generated by biology will have to be reserved for a mature and privileged group . . . this is an anti-democratic process and much to be regretted, but it seems to be the way the world is going."[15] In a capitalist system that privileged group may be corporate directors. In a socialist system it may be party bureaucrats. In both, genetic engineering provides the institutional elites with a technology far more sophisticated than any communications or weapons system presently used to secure totalitarian control.

For proponents and critics alike the debate over genetic engineering has also become a debate over the fundamental assumptions concerning the role of science and the meaning of progress.

Chief among the scientific assumptions now being contested is the right of total freedom of scientific inquiry uninhibited by any outside controls. Scientists like Edward Teller, "father of the H-bomb," believe that a researcher should not deny himself a discovery for fear of the social consequences. "I believed in the possibility of developing a thermonuclear bomb," said Teller. "My scientific duty demanded exploration of that possibility."[16]

The notion behind such statements is that scientific inquiry, if left alone, will always lead to important new insights and discoveries that will ultimately benefit humanity. To attempt to restrict the free flow of scientific inquiry is self-defeating, say the purists, since it is the only method that has proven itself capable over the years of advancing the interests of civilization. In fact, they are so convinced of this that some scientists go so far as to suggest that all of the complex issues facing our industrialized system be handed over for their profession to deal with. John Platt, research biologist at the University of Michigan, says that where problems exist, only "more scientific understanding and better technology can carry us past them." Platt argues that the escalating crisis of population expansion, hunger, nuclear proliferation, and urban decay can best be solved by a mobilization of the talent and resources of the nation's scientists, who are the only ones with the knowledge necessary to address such monumental problems.[17] Alvin Weinberg, director of the government's Oak Ridge National Laboratory, has even proposed the creation of several sociotechnical institutions to "address . . . the large social questions with much the same style and techniques that the atomic energy labs have used in developing breeders."[18]

As for the contention that science and technology are responsible for some of the critical problems facing society, the stalwarts argue that such dysfunctionalities are merely temporary nuisances which will in fact be automatically self-corrected with the passage of time and the development of new scientific technologies to superimpose on the old.

There is really no alternative to the scientific/technological approach to problem solving, argues Dr. Bernard Davis of Harvard University. "We [can't] unlearn the scientific method," says Davis, "which is available for all who wish to wrest secrets from nature."[19] In other words, science will continue to forge ahead, whether the critics like it or not.

For the first time in recent memory, there are those, even within the scientific community, who are beginning to question the assumptions underlying this line of argument. Nobel laureate Dr. James Watson argues that the notion that science, if left alone to do "its own thing," will solve everybody's problems, is "a form of laissez faire nonsense dismally reminiscent of the credo that American business, if left to itself, will solve everybody's problems. Just as the success of a corporate body in making money need not set the human condition ahead," quips Watson, "neither does every scientific advance automatically make our lives more meaningful."[20] Nobel laureate Sir Macfarlane Burnet says that the limitations of science are particularly apparent with respect to genetic engineering. "As far as medicine is concerned, molecular biology may be an evil thing. . . . It is a hard thing for an experimental biologist to accept," admits Burnet, "but it is becoming all too evident that there are dangers in knowing what should not be known."[21]

At the opening session of the National Academy of Sciences Forum on Recombinant DNA Research, Erwin Chargaff, professor emeritus of biology at Columbia University, remarked that in his long career as a teacher one of his favorite sayings to his students was "never say 'no' to an experiment." However, in the case of recombinant DNA, Dr. Chargaff told the audience of scientists, "I relinquish my old maxim" because of ethical considerations. Chargaff went on to attack the long-held supposition that scientists should follow their research wherever it leads and not be overly concerned about the effect it might have on the larger community. "I may damage myself as much as I want," said Chargaff, "but not one iota of danger to others is permissible."[22]

Some scientists, including Ethan Signer and Jonathan King of MIT and Jon Beckwith at Harvard, have become so concerned about the prospect of human genetic engineering that they have begun to speak out in public forums on the issue. "When they talk about shuffling around genes and gene therapy," says Signer, it's a sure bet that "those who are powerful in society will do the shuffling . . . The last time around, the people in charge decided the ideal genes were the ones for blue eyes, blond hair and whatever 'Aryan' meant . . ."[23] Jonathan King, a biologist whose office is on the same floor at MIT as those of some of the proponents of recombinant DNA and genetic engineering, says that he was "scared in his gut" when he began to realize what was going on down the hall in those labs. "I originally got into all of this [opposition to genetic engineering] because I considered it an immediate danger."[24] For Jon Beckwith, the Harvard biologist who led the first team to isolate a pure gene, the prospect that his work could lead to human genetic engineering was a real enough danger for him to decide not to experiment with recombinant DNA at all. "I do not wish to contribute to the development of a technology which I believe will have profound and harmful effects on this society," said Beckwith.[25] Another talented young microbiologist, Catherine Roberts, quit her promising career because of her belief that research into genetic engineering was laying the groundwork for a dangerous new eugenics policy for the country.

Speaking before the Genetic Society of America, Dr. Robert Sinsheimer, chairman of the biology department at Cal Tech, summed up much of the sentiment of the critics. "To impose any limit upon freedom of inquiry is especially bitter for the scientist, whose life is one of inquiry; but science has become too potent." Sinsheimer ended his remarks by telling his fellow geneticists that it was "no longer enough to wave the flag of Galileo."[26]

What Sinsheimer and the other scientists are saying is that not all scientific advances are progressive and that there are other ways of looking at progress besides just through the scientific perspective. The problem, says biologist Ruth Hubbard, is "that in the last 300 to 400 years the scientific method has completely overshadowed all other ways of looking at truth and reality." Hubbard claims that the scientific approach is a very narrow and limited way to look at reality, since it acknowledges as being true only that which can be measured and is repetitive. "Other kinds of things are necessary in order to really understand what is going on," argues Hubbard, "and by down-grading these other, nonscientific approaches to the world, we've gotten into a pretty bad shape."[27]

If history has taught us anything, it is that whenever one way of interpreting reality comes to be regarded as the only way to interpret reality, the conditions for enslaving the human spirit are established. This observation holds just as true for those who have blindly enshrined the scientific/technological world view as for those who have done so with other ideologies. The fact is that there are an infinite number of ways to proceed in this world and the scientific perspective is only one among them. That we have come to accept the scientific imperative over all others and have never in the past seven hundred years rejected the application of one of its major findings is in itself a testimonial to our own lack of control over our rational processes. Reason assumes the ability to choose among options. When we perceive no options, we are thinking irrationally. For too long we have failed in our responsibility to act rationally by failing to place scientific options within the context of all of the other options available to us. In the process we have narrowed our potential as human beings and done a disservice to ourselves as a species. In rejecting genetic engineering we are not rejecting progress or science. On the contrary, we are finally opening ourselves up to the realization that we have many more ways to progress than just one.

Ironically, the specter of genetic engineering has forced us to examine ourselves in toto. The very knowledge that we can now be replaced should provide a stimulus for us to prove that we are worthy of being preserved. For this reason, if for no other, it would be wrong to proceed with genetic engineering now, because it would turn us into something else right at the very moment in our history when we are ready, perhaps for the first time, to accept the responsibility of fully exploring the potential of what we are and can be.

NOTES

INTRODUCTION

1. J. William Schopf, "Evolution of Earth's Biosphere," in *The Next Billion Years: Mankind's Future in a Cosmic Perspective* (Moffett Field, Calif.: Ames Research Center, NASA, 1974), p. 26.

1 THE BIOLOGICAL REVOLUTION

1. George Wald, "The Case Against Genetic Engineering," *The Sciences,* September-October, 1976, p. 7.
2. Frederick Neidhardt, cited in "Minimizing the Risks in Genetic Engineering," *Business Week,* August 9, 1976, p. 66.

3. The Reverend Harry Hollis, interview with Randy Barber, February, 1977.
4. Liebe Cavalieri, "New Strains of Life—or Death," *New York Times Magazine,* August 22, 1976.
5. Salvador Luria, "Modern Biology: A Terrifying Power," *The Nation,* October 20, 1969, p. 406.
6. Crow cited in *BioScience,* December, 1966, p. 867.
7. *Genetic Engineering: Evolution of a Technological Issue,* Report to the Subcommittee on Science, Research and Development of the Committee on Science and Astronautics, House of Representatives, November, 1972, p. 41.
8. James D. Watson, *The Double Helix* (New York: Mentor Books, 1969), p. 126.
9. James D. Watson and Francis Crick, "Molecular Structure of Nucleic Acids," *Nature,* April 25, 1953.
10. Quoted in "Twenty-one Years of the Double Helix," *Nature,* 248 (April 26, 1974).
11. Watson, op. cit., p. 136.
12. Rick Gore, "The Awesome Worlds within a Cell," *National Geographic,* vol. 150, no. 3 (September, 1976), p. 357.
13. Lawrence Lessing, "Into the Core of Life Itself," *Fortune,* March, 1966, p. 150.
14. Charles C. Price, "Some Philosophical Implications of Evolution on the Origin and Synthesis of Life," *Perspectives in Biology and Medicine,* Spring, 1971, p. 406.
15. Robert Sinsheimer, "An Evolutionary Perspective for Genetic Engineering," *New Scientist,* January 20, 1977, p. 150.
16. James F. Danielli, "Artificial Synthesis of New Life Forms," *Bulletin of Atomic Scientists,* December, 1972, p. 21.
17. Jacques Yves Cousteau, "An Artificial Ocean for

an Artificial Planet," in *The Next Billion Years: Mankind's Future in a Cosmic Perspective* (Moffett Field, Calif.: Ames Research Center, NASA, 1974), p. 49.

18. James Bonner, "Beyond Man's Genetic Lottery," in *The Next Billion Years,* op. cit., p. 59.
19. Joseph Fletcher, "Ethical Aspects of Genetic Controls," *New England Journal of Medicine,* September 30, 1971, p. 782.
20. Harlow Shapley, cited in Albert Rosenfeld, *The Second Genesis* (New York: Vintage Press, 1969), p. 165.
21. Joseph Fletcher, *The Ethics of Genetic Control* (Garden City, N.Y.: Anchor Books, 1974), p. xvii.
22. Sydney Brenner, "New Directions in Molecular Biology," *Nature,* 248 (April 26, 1974), p. 786.
23. Ibid.
24. Robert L. Sinsheimer, "Genetic Engineering: The Modification of Man," *Impact of Science on Society,* XX (1970), p. 288.
25. Brenner, op. cit., p. 786.
26. *Genetic Engineering: Evolution of a Technological Issue,* op. cit., Supplemental Report I, December, 1974, p. 5.
27. Lessing, op. cit., p. 150.
28. Ibid., p. 152.
29. Ibid.
30. Joel N. Shurkin, "Yet Another Step in the Complex Probe of the Genetic Code," *Philadelphia Inquirer,* May 1, 1977.
31. *Newsweek,* June 1, 1970, p. 46.
32. Graham Chedd, "Danielli the prophet," *New Scientist and Science Journal,* January 21, 1971, p. 124.
33. D. S. Halacy, Jr., *The Genetic Revolution* (New York: Harper & Row, 1974), p. 185.
34. *DNA Recombinant Molecule Research,* Supple-

mental Report II, Report prepared for the Subcommittee on Science, Research and Technology of the Committee on Science and Technology, House of Representatives, December, 1976, p. 16.

35. Jonathan Beckwith, "Gene Expression in Bacteria and Some Concerns About the Misuse of Science," *Bacteriological Review,* vol. 34, no. 3 (September, 1970), p. 224.
36. Jonathan Beckwith, "Recombinant DNA: Does the Fault Lie Within Our Genes?", paper presented to the National Academy of Sciences Forum on Recombinant DNA, Washington, D.C., March 7–9, 1977.
37. *New York Times,* April 24, 1977, Section 4, p. 6.
38. "A Working Synthetic Gene," *Medical World News,* September 20, 1976, p. 7.
39. C. R. Merril et al., "Bacterial versus Gene Expression in Human Cells," *Nature,* 233, pp. 398–400.
40. Personal interview with Dr. Ethan Singer, April, 1977.
41. Cited in Gore, op. cit., p. 385.
42. Brenner, op. cit., p. 785.
43. Ibid., p. 787.
44. Marshall Nirenberg, statement by the National Institutes of Health, 1970.
45. Lord Ritchie-Calder, "Retailoring the Tailor," *1976 Encyclopedia Britannica, Book of the Year,* Special Supplement, p. iv.
46. Luria, op. cit., p. 408.
47. "Genetics: Conference Sets Strict Controls to Replace Moratorium," *Science,* March 14, 1975, p. 932.
48. Robert Graves, *New Larousse Encyclopedia of Mythology* (New York: Prometheus Press, 1971), p. 146.
49. Stanley Cohen, "The Manipulation of Genes,"

Scientific American, vol. 233, no. 1 (July, 1975), p. 25.

50. Cited in Ronald Kotulak, "The Great Promise and Grave Peril of Genetic Manipulation," *Chicago Tribune Magazine,* September 21, 1975, p. 30.
51. Dr. Sydney Brenner, cited in Janet Weinberg, "Asilomar Decision: Unprecedented Guidelines for Gene-Transplant Research," *Science News,* 107 (March 8, 1975), p. 148.
52. Oliver Smithies, letter to Dr. DeWitt Stetten, Jr., November 26, 1976. Cited in *DNA Recombinant Molecule Research,* Supplemental Report II, op. cit., p. 6.
53. Victor Cohn, "Gene Tied to Growth Hormone by California Researchers," *Washington Post,* May, 1977.
54. Cohen, op. cit., p. 32.
55. Joshua Lederberg, "DNA Splicing: Will Fear Rob Us of Its Benefits?", *Prism,* November, 1975, p. 33.
56. Ibid., p. 37.
57. Cited in M. Yao, "Scientists Split on DNA Research," *The Michigan Daily,* March 4, 1976, p. 1.
58. Bernard N. Davis, "Darwin, Pasteur, and the Andromeda Strain," Public Lecture, Harvard Science Center, reproduced in *DNA Recombinant Molecule Research,* Supplemental Report II, op. cit., p. 253.
59. Ibid.
60. Jeremy Rifkin, Larry Gordon, and Dan Smith, "DNA," *Mother Jones,* February-March, 1977, p. 23.
61. *Fortune,* February, 1974, p. 100.
62. Max Birnstiel, cited in "Genetic Engineering and Clashing Views," *Science News,* 106 (November 2, 1974), p. 277.

63. Berg cited in Frances R. Warshaw, *Gene Implantation: Proceed with Caution* (Boston: Science for the People), November, 1976, p. 28.
64. Liebe Cavalieri, "New Strains of Life or Death?", *New York Times Magazine,* August 22, 1976.
65. Ibid.
66. Ibid.
67. Paul Berg et al., "Potential Biohazards of Recombinant DNA Molecules," *Science,* 185 (July 26, 1974).
68. William Bennett and Joel Gurin, "Science that Frightens Scientists," *Atlantic Monthly,* February, 1977, p. 49.
69. Cited in Robert Kanigel, "Pandora's Box, Chapter XI: Splicing the Double Helix," *Johns Hopkins Magazine,* November, 1976, p. 30.
70. Erwin Chargaff, "A Slap at the Bishops of Asilomar," *Science,* vol. 190 (October 10, 1975), p. 135.
71. Donald Brown, testimony, *Genetic Engineering, 1975,* before the Subcommittee on Health, Committee on Labor and Public Welfare, U.S. Senate, April 22, 1975, p. 24.
72. Jonathan King, cited in Bennett and Gurin, op. cit., p. 56.
73. Janet H. Weinberg, "Decision at Asilomar," *Science News,* 107 (March 22, 1975), p. 196.
74. Michael Rogers, "The Pandora's Box Congress," *Rolling Stone,* no. 189 (June 19, 1975), p. 77.
75. "Genetic Conference Sets Strict Controls to Replace Moratorium," *Science,* 187 (March 14, 1975), p. 932.
76. Jonathan King, testimony before the Cambridge City Council, June 23, 1976.
77. Ibid.
78. Ruth Hubbard, "Gazing Into the Crystal Ball," *Bio-Science,* vol. 26, no. 10 (October, 1976), p. 608.

79. Dr. Willard Gaylin, testimony, *Genetic Engineering, 1975,* op. cit.
80. Sinsheimer, "An Evolutionary Perspective for Genetic Engineering," op. cit., p. 151.
81. Cited in Kanigel, op. cit., p. 29.
82. Lederberg, op. cit., p. 36.
83. Arthur Kornberg, letter to Cambridge City Council, cited in Wald, op. cit., p. 8.
84. Cited in "Minimizing the Risks in Genetic Engineering," *Business Week,* August 9, 1976, p. 67.
85. Cited in Kanigel, op. cit., p. 32.
86. Donald Frederickson, *Federal Register,* vol. 41, no. 176 (September 9, 1976), p. 38,435.
87. Paul Berg, "Recombinant DNA Research Must and Can be Done Safely," *Trends in Biochemical Sciences,* February, 1977.
88. Maxine F. Singer, "The Recombinant DNA Debate," *Science,* 196 (April 8, 1977), p. 127.
89. Sinsheimer, "An Evolutionary Perspective for Genetic Engineering," op. cit., p. 152.
90. Ibid., p. 151.
91. Dr. Roger Lewin, "Genetic Engineers Ready for Stage Two," *New Scientist,* October 14, 1976, p. 85.
92. Ibid., p. 86.

2
EUGENICS: THE IDEOLOGY BEHIND GENETICS RESEARCH

1. Theodore Roosevelt to Charles B. Davenport, January 3, 1913, Charles B. Davenport Papers, Department of Genetics, Cold Spring Harbor, N.Y. Roosevelt "Birth Reform, from the Positive, Not the Negative Side," in *The Works of Theodore Roosevelt* (New York: Charles Scribner's Sons, 1923–26), XXI, p. 163.

2. Roosevelt, "Twisted Eugenics," in *The Works of Theodore Roosevelt,* op. cit., National Edition, XII, p. 201.
3. *American Eugenics Proceedings of the Annual Meeting,* American Eugenics Society, 1936.
4. Garrett Hardin, *Nature and Man's Fate* (New York: Rinehart and Co., 1959), p. 217. Also: Karl Pearson, *The Life, Letters and Labours of Francis Galton* (Cambridge University Press, 1930).
5. Hardin, op. cit., p. 216.
6. Francis Galton, *Hereditary Genius: An Inquiry Into Its Laws and Consequences* (New York: World Publishing Co., 1962).
7. Ibid.
8. Noyes, *Essay on Scientific Propagation* (Oneida, N.Y.), pp. 2, 5.
9. Noyes, "The Oneida Community Experiment in Stirpiculture," *Eugenics, Genetics and the Family,* Scientific Papers of the 2nd International Congress of Eugenics (Baltimore: The Wilkans and Wilkans Co., 1923).
10. Pearson, op. cit., vol. III A, p. 435.
11. Galton, *Inquiries Into Human Faculty and Its Development* (New York: E. P. Dutton and Co., 1908), p. 200.
12. K. M. Ludmerer, *Genetics and American Society* (Baltimore: Johns Hopkins University Press, 1972), p. 38.
13. Ibid., p. 39.
14. Ibid.
15. Ibid., p. 43.
16. Ibid.
17. Michael F. Guyer, *Being Well Born* (Indianapolis: Bobbs-Merrill Co., 1916), preface.
18. E. G. Conklin, "The Future of America: A Biological Forecast," *Harper's Magazine,* 156, 1928. (Written anonymously.)

19. H. H. Newman, *Evolution, Genetics and Eugenics* (Chicago: University of Chicago Press, 1921), p. 441.
20. American Breeders Association *Proceedings,* II (1906).
21. Quotation from Davenport, Charles B. Davenport Papers, op. cit.
22. Mark H. Haller, *Eugenics* (Rutgers University Press, 1963), p. 73.
23. Davenport Papers, July 22, 1913.
24. R. B. Von Kleinsmid, "An Inquiry Concerning Some Preventions of Crime," *APA* (1915), p. 108.
25. McDougall, *Is America Safe for Democracy?* (New York: Charles Scribner's Sons, 1921); *Ethics and Some Modern World Problems* (New York: G. P. Putnam's Sons, 1924); *The Indistinctible Union* (Boston: Little, Brown and Co., 1925).
26. Quoted in *Teaching School Bulletin,* X (February, 1914), p. 160.
27. Statement by Irving Fisher, "What I Think About Eugenics," in Eugenics Society of the U.S.A., *A Brief Bibliography of Eugenics,* p. 5.
28. "IQ: Scientific or Social Controversy," *Science for the People*, March, 1974.
29. Earnest Albert Hooton, *The American Criminal* (Cambridge: Harvard University Press, 1939), pp. 307–309.
30. Jon Beckwith, "Social and Political Uses of Genetics in the U.S.: Past and Present," *Annals of the New York Academy of Science,* 265 (1976), p. 47.
31. E. Godkin, *Problems of Modern Democracy, Political and Economic Essays* (Cambridge: Belknap Press of Harvard University Press, 1966).

32. H. Croly, *The Promise of American Life* (New York: Macmillan Co., 1909), p. 81.
33. Calvin Coolidge, "Whose Country is This?", *Good Housekeeping,* 72 (February, 1921).
34. "A Few Thoughts Concerning Eugenics" by A. G. Bell (an address to the American Breeders Association in Washington, January 1908), *National Geographic,* February, 1908. Also: Bell, "How to Improve the Race," *Journal of Heredity,* January, 1914.
35. Sanger, "Need for Birth Control in America," in Adolf Meyer, ed., *Birth Control, Facts and Responsibilities* (Baltimore: Williams and Williams Co., 1925), p. 15.
36. Jordan, *The Human Harvest* (Boston: American Unitarian Association, 1910).
37. David Starr Jordan, *The Blood of the Nation* (Boston: American Unitarian Association, 1902).
38. Caleb Williams Saleeby, *The Progress of Eugenics* (New York: Funk & Wagnalls, 1914), p. 89.
39. David S. Jordan to Davenport, February 3, 1913, Davenport Papers; *Eugenical News,* VII (October, 1922), p. 111; and "Fitter Families Again," *Journal of Heredity,* XVII (February, 1926), p. 68–69.
40. *Criminal Man* (New York: G. P. Putnam's Sons, 1911), p. 7; and H. Ellis, *The Criminal* (London: Walter Scott, Ltd., 1897), p. 83.
41. Ibid.
42. H. M. Boies, *Prisoners and Paupers* (Freeport, N.Y.: Books for Libraries Press, 1972).
43. W. D. McKim, *Heredity and Human Progress* (New York: G. P. Putnam's Sons, 1900).
44. M. W. Barr, *Mental Defectives* (Philadelphia: P. Blakiston's Sons and Co., 1904).
45. Sharp, "Rendering Sterile of Confirmed Criminals and Mental Defectives," *Proceedings of the National Prison Association* (1907), p. 178.

46. Ibid., p. 180; Laughlin, *Eugenical Sterilization in the United States* (Chicago: Psychopathic Laboratory of the Court of Chic), p. 15.
47. H. B. Hickman, "Delinquent and Criminal Boys Tested by the Binet Scale," *Teaching School Bulletin,* XI (January, 1915), p. 159.
48. Statement in *Report of the State Commission to Investigate Provisions for the Mentally Deficient* (Albany: J. B. Lyon Co., 1915), p. 178.
49. Henry H. Goddard, *The Kallikak Family* (New York: Macmillan Co., 1912).
50. Haller, op. cit., p. 108.
51. Harry H. Laughlin, *Scope of the Committee's Work* (Cold Spring Harbor, N.Y.: Eugenics Record Office, 1914).
52. J. H. Landman, *Human Sterilization* (New York: Macmillan Co., 1932), p. 259.
53. Haller, op. cit., p. 139.
54. Landman, op. cit., pp. 80–93.
55. R. M. Yerkes, *Psychological Examining in the U.S. Army,* National Academy of Sciences, Memoirs XV (Washington, D.C., 1912). Also: Frederick Osborn and Frank Lorimer, *Dynamics of Population* (New York: Macmillan Co., 1934).
56. Goddard, op. cit., pp. 78, 90.
57. G. H. Hardy, "Mendelian Propositions in a Mixed Population," *Science,* 28 (1908), pp. 49–50. Also: Edward East, "A Mendelian Interpretation of Variation that is Apparently Continuous," *American Naturalist,* 44 (1910), pp. 65–82. And R. A. Emerson and E. M. East, "The Inheritance of Qualitative Characters in Maize," *Bulletin of the Agricultural Experimental Station of Nebraska* (1913), p. 120.
58. "Eugenics, Genetics & Family," *Science Papers of the Second International Congress of Eugenics,* 1 (Baltimore: Williams & Williams Co., 1923).

59. Fisher to Davenport, March 2, 1912, Davenport Papers.
60. Ludmerer, op. cit., p. 25.
61. Ross, *The Old World in the New* (New York: Century Co., 1944), pp. 113, 145, 147, 148, 150.
62. Madison Grant, *The Conquest of a Continent* (New York: Charles Scribner's Sons, 1933), p. 53; *Passing of the Great Race* (New York: Charles Scribner's Sons, 1916), p. 78.
63. Anthony Smith, *The Human Pedigree* (New York: J. B. Lippincott Company, 1975), p. 233.
64. Robert Divine, *American Immigration Policy* (New Haven: Yale University Press, 1957), pp. 5–18.
65. Garland E. Allen, "Genetics, Eugenics and Class Struggle," *Genetics,* June, 1975.
66. Harry H. Laughlin, "Analysis of America's Melting Pot," Hearings before the House Committee on Immigration and Naturalization, 67th Congress, 3rd session (U.S. Government Printing Office, 1922), p. 755.
67. James J. Davis, "Our Labor Shortage and Immigration," *Industrial Management,* 65 (1923), p. 323.
68. *Congressional Record,* April 8, 1924, p. 5693.
69. Ibid., p. 5872.
70. "Restriction and Immigration," Hearings before House Committee on Immigration and Naturalization, 68th Congress, 1st Session (U.S. Government Printing Office, 1924), pp. 767–768.
71. Newman, *Evolution, Genetics and Eugenics* (Chicago: University of Chicago Press, 1932 edition), p. 531.
72. Beckwith, *Social and Political Uses of Genetics,* op. cit., p. 49.
73. Adolf Hitler, *Mein Kampf* (Boston: Houghton Mifflin Company, 1943 edition).

74. L. C. Dunn, "Cross Currents in the History of Human Genetics," *American Journal of Human Genetics,* 14 (1962), p. 8.
75. Paul Popenoe, "The German Sterilization Law," *Journal of Heredity,* 25 (1934).
76. Conversation with L. C. Dunn, October 5, 1970, in K. M. Ludmerer, op. cit., p. 133.
77. Ibid., p. 118.
78. U.S. Atomic Energy Commission, *Atomic Energy and the Life Sciences* (Washington, D.C.: U.S. Government Printing Office, 1949), p. 187.
79. D. S. Halacy, Jr., *Genetic Revolution* (New York: Harper & Row, 1974), p. 93.
80. Frederick Ausubel, Jon Beckwith, and Karen Janssen, "The Politics of Genetic Engineering: Who Decides Who's Defective?" *Psychology Today*, June, 1974.
81. Anthony Smith, op. cit., p. 247.
82. Ibid., p. 221.
83. Interview with Dr. Samenow, *60 Minutes,* February 20, 1977.
84. Ibid.
85. David Rorvik, "Behavior Control; Big Brother Games," *Intellectual Digest,* January, 1974.
86. Ibid.
87. Bentley Glass, "Science: Endless Horizons or Golden Age?", *Science,* 171 (January 8, 1971), p. 23.
88. "IQ: Scientific or Social Controversy," *Science for the People,* February 1976, pp. 39–40.
89. Ibid., p. 42.
90. Ibid., pp. 27–36.
91. *Daily Californian,* April 2, 1973.
92. Charles Frankel, "The Specter of Eugenics," *Commentary,* March, 1974.
93. Beckwith, "Social and Political Uses of Genetics," op. cit., p. 51.

94. Rose, Hambley, and Haywood, *The Socialist Register* (London: Merlin Press, 1974), p. 236.
95. Ausubel, Beckwith, and Janssen, op. cit., p. 40.
96. Ibid., p. 30.
97. Jon Beckwith, "Social and Political Uses of Genetics in the U.S.: Past and Present," op. cit., p. 53.
98. W. T. Vukovich, "The Dawning of the Brave New World," *University of Illinois Law Forum,* 1971, pp. 189–231.
99. Ibid.
100. Charles Frankel, op. cit.
101. Ibid.
102. Ausubel, Beckwith, and Janssen, op. cit.

3
LIFE IN THE LABORATORY

1. Lord Ritchie-Calder, "The Tailor Retailored," *1976 Britannica Book of the Year,* Special Supplement, p. iv.
2. Julian Huxley cited in Joseph Fletcher, *The Ethics of Genetic Control* (Garden City, N.Y.: Anchor Books, 1974), p. 76.
3. Joshua Lederberg, "Experimental Genetics and Human Evolution," *Bulletin of the Atomic Scientists,* October, 1966, p. 6.
4. Donald Fleming, "On Living in a Biological Revolution," *Atlantic,* vol. 223, no. 2 (February, 1969), p. 68.
5. Personal interview with George Wald, April, 1977.
6. Leon Kass, "Making Babies—The New Biology and the 'Old' Morality," *The Public Interest,* Winter, 1972, p. 48.
7. Jacques Ellul, *The Technological Society* (New York: Vintage, 1974), p. 11.

8. Salvador Luria, "Modern Biology: A Terrifying Power," *The Nation,* October 20, 1969, p. 408.
9. Joseph Fletcher, *The Ethics of Genetic Control* (Garden City, N.Y.: Anchor Books, 1974), p. 15.
10. Kerstin Aner, "Genetic Manipulation as a Political Issue," in Charles Birch and Paul Abrecht, eds., *Genetics and the Quality of Life* (Sydney: Pergamon Press, 1975).
11. Joshua Lederberg, "Biological Innovation and Genetic Intervention," in John A. Behnke, ed., *Challenging Biological Problems: Directions Toward Their Solution* (1972), pp. 16–17.
12. Joshua Lederberg, testimony, *Hearings,* Committee on Government Operations, U.S. Senate, Joint Res. 145, March 7–28, April 2, 1968, p. 67.
13. Fletcher, op. cit.
14. *Genetic Engineering: Evolution of a Technological Issue,* Report to the Subcommittee on Science, Research and Development of the Committee on Science and Astronautics, November, 1972, p. 26.
15. "Genetic Engineering: Reprise," *Journal of the American Medical Association,* vol. 220, no. 10 (June 5, 1972), pp. 1356–57.
16. Leon Kass, "New Beginnings of Life," in Michael Hamilton, ed., *The New Genetics and the Future of Man* (Grand Rapids, Mich.: Eerdmans), pp. 15–63.
17. Joseph Fletcher, "Ethical Aspects of Genetic Controls," *New England Journal of Medicine,* 285 (1971), p. 776.
18. Boyce Rensberger, "2 Banks for Freezing Human Sperm Planned in Midtown, With Services Available to Public," *New York Times,* August 18, 1971, p. 19.

19. Edward Grossman, "The Obsolescent Mother: A Scenario," *Atlantic,* May, 1971, p. 47.
20. D. S. Halacy, Jr., *Genetic Revolution* (New York: Harper & Row, 1974), p. 149.
21. Quoted in Jane E. Brody, "1976 Marks Tricentennial of the Discovery of Sperm," *New York Times,* December 20, 1975, p. 29.
22. Gordon Leach, *The Biocrats* (Great Britain: Penguin Books, 1972), p. 86.
23. Robert T. Francoeur, "We *Can*—We Must: Reflections on the Technological Imperative," *Theological Studies,* vol. 33, no. 3 (September, 1972), p. 431.
24. Ibid.
25. Halacy, op. cit., p. 148.
26. Albert Rosenfeld, *The Second Genesis* (New York: Vintage, 1969), p. 175.
27. Cited in Ethan Jacobson, "Up 400%: Artificial Insemination," *Hospital Tribune,* December 6, 1976, p. 9.
28. John Olsen, interview with Dan Smith, May 19, 1977.
29. Constance Holden, "Sperm Banks Multiply as Vasectomies Gain Popularity," *Science,* vol. 176 (April 7, 1972), p. 32.
30. Cited in Boyce Rensberger, op. cit., p. 19.
31. John Olsen, op. cit.
32. Ethan Jacobson, op. cit., p. 9.
33. Ibid., p. 12.
34. Cited in Constance Holden, op. cit., p. 32.
35. *Genetic Engineering,* op. cit., p. 17.
36. H. J. Muller, "Human Evolution by Voluntary Choice of Germ Plasm," *Science,* vol. 134 (1961), pp. 643–49.
37. Caryl Rivers, "Genetic Engineering: Now That They've Gone Too Far, Can They Stop?" *Ms.,* June 1976, p. 112.
38. Warren B. Miller, "Reproduction, Technology

and the Behavioral Sciences," *Science,* vol. 183, no. 4121 (January 18, 1974), p. 149.
39. Rivers, op. cit., p. 112.
40. "The Riddle of A.I.," *Time,* February 25, 1966, p. 48.
41. Ibid.
42. Albert Rosenfeld, "Will Man Direct His Own Evolution?", *Life,* vol. 59, no. 14 (October 1, 1965).
43. Jean L. Marx, "Embryology: Out of the Womb—into the Test Tube," *Science,* vol. 182 (November 23, 1973), p. 814.
44. *Los Angeles Times,* June 16, 1971, section 3, p. 1.
45. Walter Sullivan, "Frozen Mouse Embryos Grow in Foster Mothers," *New York Times,* August 15, 1972, p. 16.
46. Halacy, op. cit., p. 152.
47. "Bullseye," *Nature,* 243 (June 15, 1973), p. 371.
48. Jean Marx, op. cit., p. 811.
49. Cited by David Rorvik, "The Embryo Sweepstakes," *New York Times Magazine,* September 15, 1974, pp. 16–17.
50. "Human Ova Fertilized After Frozen Storage," *Medical World News,* March 5, 1965, p. 34.
51. Albert Rosenfeld, *The Second Genesis,* op. cit., p. 181.
52. James D. Watson, "The Future of Asexual Reproduction," *Intellectual Digest,* vol. 2, no. 2 (1971), pp. 69–74.
53. Albert Rosenfeld, *Second Genesis,* op. cit., p. 41.
54. Rorvik, op. cit., p. 50.
55. Edward Grossman, op. cit., p. 43.
56. Ibid.
57. Ibid.
58. Gerald George, "Creating Life in the Lab," *Science Digest,* January, 1974, p. 29.

59. Kenneth Guentert, "Will Your Grandchild Be a Test-Tube Baby?", *U.S. Catholic,* June, 1977, p. 9.
60. Rorvik, op. cit., p. 55.
61. Ibid.
62. Cited in Gerald George, op. cit., p. 28.
63. Rorvik, op. cit., p. 55.
64. "Unit Proposed for in vitro Fertilization," *Nature,* 245 (September 7, 1973), p. 3.
65. R. G. Edwards and R. F. Fowler, "Human Embryos in the Laboratory," *Scientific American,* 233 (December, 1970), p. 54.
66. Cited in Gerald George, op. cit., p. 33.
67. "Invit: The View from the Glass Oviduct," *Saturday Review/Science,* September 30, 1972, p. 202.
68. Bentley Glass, "Science: Endless Horizons or Golden Age?", *Science,* vol. 171, no. 3966 (January 8, 1971), p. 28.
69. Leon Kass, "Babies by Means of In Vitro Fertilization: Unethical Experiments on the Unborn?", *New England Journal of Medicine,* November 18, 1972.
70. Joseph Fletcher, *The Ethics of Genetic Control,* op. cit., p. 103.
71. *Genetic Engineering,* op. cit., p. 20.
72. Cited in Rosenfeld, *Second Genesis,* op. cit., p. 139.
73. Jean Marx, op. cit., p. 811.
74. "Control of Life," *Life,* September 10, 1965, p. 76.
75. Kenneth Guentert, op. cit., p. 10.
76. Caryl Rivers, "Genetic Engineering Portends a Grave New World," *Saturday Review,* April 8, 1972, pp. 23–27.
77. Edward Grossman, op. cit., p. 45.
78. "Control of Life," op. cit., p. 62.
79. Ibid., p. 67.

80. *Genetic Engineering,* op. cit., p. 34.
81. Edward Grossman, op. cit., p. 48.
82. Joseph Fletcher, "Ethical Aspects of Genetic Controls," *New England Journal of Medicine,* September 30, 1971.
83. Cited in Rorvik, op. cit., p. 16.
84. Aldous Huxley, *Brave New World* (New York: Harper & Row, 1969).
85. James D. Watson, "Moving Toward the Clonal Man," *Atlantic,* vol. 227, no. 5 (May, 1971), p. 52.
86. Joshua Lederberg, "Experimental Genetics on Human Evolution," op. cit., p. 9.
87. "A Turkey that Never had a Father," *Life,* April 16, 1956, p. 111.
88. Halacy, op. cit., p. 160.
89. Ibid.
90. Henry Still, *Man-Made Men* (New York: Hawthorn Books, 1973), p. 217.
91. Cited in David Rorvik, "Cloning: Asexual Human Reproduction," *Science Digest,* November, 1969, p. 9.
92. Rick Gore, "The Awesome Worlds within a Cell," *National Geographic,* vol. 150, no. 3 (September, 1976), p. 388.
93. Halacy, op. cit., p. 160.
94. Henry Still, *Man-Made Men* (New York: Hawthorn Books, 1973).
95. Rorvik, "Cloning: Asexual Human Reproduction," op. cit., p. 9.
96. Watson, "The Future of Asexual Reproduction," op. cit.
97. John R. Platt, *Science,* vol. 154 (December 2, 1966), p. 1135.
98. Lederberg, "Experimental Genetics and Human Evolution," op. cit., p. 9.
99. Cited in Rorvik, "Cloning: Asexual Human Reproduction," op. cit., p. 7.

100. Rivers, "Genetic Engineering: Now That They've Gone Too Far, Can They Stop?", op. cit., p. 51.
101. Halacy, op. cit., p. 163.
102. Lederberg, "Experimental Genetics and Human Evolution," op. cit., p. 10.
103. James Danielli, "Artificial Synthesis of New Life Forms," *Bulletin of Atomic Scientists,* December, 1972, p. 24.
104. Lederberg, op. cit., p. 10.
105. Cited in Rosenfeld, *Second Genesis,* op. cit., p. 149.
106. Lederberg, op. cit., p. 10.
107. J. B. S. Haldane, "Biological Possibilities in the Next Ten Thousand Years," in Gordon Wolstenholme, ed., *Man and His Future* (Boston: Little, Brown & Company, 1963), p. 352.
108. Bernard Davis, "Prospect for Genetic Intervention in Man," *Science,* 170 (December 18, 1970), p. 1282.
109. *Los Angeles Times,* May 17, 1971, part iv, p. 1.
110. Gunther Stent, "Molecular Biology and Metaphysics," *Nature,* 248 (April 26, 1974).
111. Cited in Rosenfeld, *Second Genesis,* op. cit., p. 151.
112. *Time,* April 19, 1971, pp. 33 ff.
113. Ibid.
114. Bernard Davis, "Gene Engineering: How Great is the Danger?", *Science,* vol. 186, no. 4161 (October 25, 1974), p. 7.
115. Joshua Lederberg, "Genetic Engineering and the Amelioration of Genetic Defects," *BioScience,* 20, 24 (December 15, 1970), p. 1307.
116. *Genetic Engineering,* op. cit., p. 22.
117. Lederberg, "Experimental Genetics and Human Evolution," op. cit., p. 10.
118. I. Scott Bass, "Government Control of Research in Positive Eugenics," *University of Michigan Journal of Law,* Spring, 1974, p. 621.

4
ELIMINATING THE "BAD" GENES

1. Cited in Albert Rosenfeld, *The Second Genesis* (New York: Vintage, 1975), p. 179.
2. Cited in *Los Angeles Times,* November 23, 1969, p. 1.
3. Donald Huisingh, "Should Man Control His Genetic Future?", *Zygon,* 42 (February, 1969), p. 198.
4. National Institute of General Medical Sciences, *What Are the Facts About Genetic Disease,* (DHEW Publication No. NIH 75-370), p. 6.
5. Cited in Caryl Rivers, "Genetic Engineering: Now That They've Gone Too Far, Can They Stop?", *Ms.,* June, 1976, p. 11.
6. Cited in D. S. Halacy, Jr., *Genetic Revolution* (New York: Harper & Row, 1974), p. 101.
7. Caryl Rivers, "Genetic Engineering Portends a Grave New World," *Saturday Review,* April 8, 1972, pp. 23–27.
8. Gene Bylinsky, "What Science Can Do About Hereditary Diseases," *Fortune,* September, 1974, p. 156.
9. *Genetic Engineering: Evolution of a Technological Issue,* Report to the Subcommittee on Science, Research and Development of the Committee on Science and Astronautics, U.S. House of Representatives, November, 1972, p. 43.
10. Paul T. Libassi, "Biochemical and Chromosomal Defects Yield to Prenatal Diagnosis," *Laboratory Management,* vol. 10, no. 9 (1972), pp. 20–24.
11. Richard Restak, "The Danger of Knowing Too Much," *Psychology Today,* September, 1975, p. 92.
12. Bylinsky, op. cit., p. 152.
13. Ibid., p. 149.
14. Libassi, op. cit.

15. Barbara J. Culliton, "Genetic Screening: States May Be Writing the Wrong Kinds of Laws," *Science,* 191 (March 5, 1976), p. 926.
16. Lucy Eisenberg, "Genetics and the Survival of the Unfit," *Harper's,* February, 1966, p. 54.
17. Culliton, op. cit., p. 926.
18. Ibid.
19. Ibid.
20. Ibid.
21. Bylinski, op. cit., p. 158.
22. Jonathan King, Jon Beckwith, and Larry Miller, "Genetic Screening: Pitfalls," *The Science Teacher,* 43, 5 (May, 1976).
23. Paul Ramsey, "Screening: An Ethicist's View," *Ethical Issues in Genetic Engineering,* Bruce Hilton et al., eds. (New York: Plenum Press, 1973), p. 159.
24. Restak, op. cit., p. 22.
25. Ibid., p. 92.
26. Ibid., p. 88.
27. "Bioethics: A Mounting Dilemma for Geneticists," *Laboratory Management,* vol. 10, no. 11 (1972), pp. 34–37.
28. Restak, op. cit., p. 22.
29. Bentley Glass, "Science: Endless Horizons or Golden Age," *Science,* 171 (January 8, 1971), p. 28.
30. Eisenberg, op. cit., p. 53.
31. Cited in Walter G. Peter, III, "Ethical Perspectives in the Use of Genetic Knowledge," *BioScience,* 21 (1971), pp. 1133–37.
32. "Bioethics: A Mounting Dilemma for Geneticists," op. cit., p. 168.
33. King, Beckwith, Miller, op. cit.
34. Cited in Walter G. Peter, III, op. cit., p. 205.
35. S. E. Luria, "Modern Biology: A Terrifying Power," *The Nation,* October 20, 1969, p. 408.

36. Darrel S. English, "Genetic Manipulation and Man," in Darrel English, ed., *Genetic and Reproductive Engineering* (New York: MSS Information Corp., 1974), p. 58.
37. George A. Hucock, "Gene Therapy and Genetic Engineering: Frankenstein is Still a Myth, But It Should Be Read Periodically," *Indiana Law Journal,* vol. 48, no. 4 (Summer, 1973), p. 544.
38. English, op. cit., p. 59.
39. Robert Pollack, "Tumors and Embryogenesis," *Science,* 194, p. 1272.
40. Caryl Rivers in *Ms.,* op. cit., p. 116.
41. Cited in Victor Cohn, "Scientists Duplicate Rat Insulin Gene," *Washington Post,* May 23, 1977, p. 1.
42. National Institute of General Medical Sciences, *Annual Report, FY 1976,* p. 24.
43. Paul T. Libassi, "Getting Down To Business in Genetics: Research, Diagnosis, Treatment," *Laboratory Management,* vol. 10, no. 8 (1972), pp. 24–30.
44. *What Are the Facts About Genetic Disease,* op. cit., p. 29.
45. Ibid., p. 4.
46. Libassi, "Getting Down to Business in Genetics," op. cit., p. 49.
47. Theodore Friedmann, "The Future for Gene Therapy—A Reevaluation," *Annals of the New York Academy of Sciences,* 1975, p. 141.
48. H. Vasken Aposhian, "The Use of DNA for Gene Therapy—The Need, Experimental Approach, and Implications," *Perspectives in Biology and Medicine,* vol. 14, no. 1 (Autumn, 1970), p. 108.
49. Robert Sinsheimer, "Genetic Engineering: The Modification of Man," *Impact of Science on Society,* XX (1970), p. 286.

50. Kenneth J. McDougall, "Genetic Engineering: Hazard or Blessing?", *Intellect,* April, 1976, p. 529.
51. Aposhian, op. cit., p. 107.
52. Theodore Friedmann and Richard Roblin, "Gene Therapy for Human Genetic Disease?", *Science,* March 3, 1972, pp. 953, 954.
53. Huisingh, op. cit.
54. Marc Lappé, "Moral Obligations and the Fallacies of Genetic Control," *Theological Studies,* vol. 33, no. 3 (September, 1972), p. 422.
55. Cited in Rosenfeld, *Second Genesis,* op. cit., p. 212.
56. Lord Ritchie-Calder, "The Tailor Re-tailored," *1976 Britannica Book of the Year,* Special Supplement, p. v.
57. Julian Huxley, "The Future of Man—Evolutionary Aspects," in Gordon Wolstenholme, ed., *Man and His Future* (Boston: Little, Brown, 1963), p. 12.
58. Robert Reinhold, "Challenge of Genetics Fades for Scientists," *New York Times,* August 27, 1973, p. 34.
59. W. French Anderson, "Genetic Therapy," in Michael P. Hamilton, ed., *The New Genetics and the Future of Man* (Grand Rapids: Eerdmans, 1972), p. 119.
60. Bernard Davis, "Prospects for Genetic Intervention in Man," *Science,* 170 (December 18, 1970), p. 1281.
61. Charles Frankel, "The Specter of Eugenics," *Commentary,* 57 (March, 1974), p. 31.
62. Lederberg, "Experimental Genetics and Human Evolution," op. cit., p. 7.
63. James J. Nagle, "Genetic Engineering," *Bulletin of the Atomic Scientists,* December 1971, p. 44.
64. Lederberg, "Experimental Genetics and Human Evolution," op. cit., p. 9.

65. Glass, op. cit., p. 28.
66. Cited in Rosenfeld, *Second Genesis,* op. cit., p. 33.
67. "Man into Superman: The Promise and Peril of the New Genetics," *Time,* April 19, 1971.
68. "Genetic Screening: Pitfalls," op. cit.
69. Russell Baker, "Dr. Irving Slezak: One Very Smart Tomato," *Lincoln Star,* May 4, 1977.
70. Cited in Rosenfeld, *Second Genesis,* op. cit., p. 140.
71. Robert O. Becker, "Stimulation of Partial Limb Regeneration in Rats," *Nature,* 235, (January 14, 1972), p. 109.
72. Cited in Albert Rosenfeld, "Will Man Direct His Own Evolution?", *Life,* October 1, 1975, p. 105.
73. Dr. James Bonner, in *The Far Reach of Science,* (California Institute of Technology, 1968).
74. Lawrence Lessing, "At the Controls of the Living Cell," *Fortune,* May, 1966, p. 206.
75. Cited in D. S. Halacy, Jr., op. cit., p. 136.
76. Lederberg, "Experimental Genetics and Human Evolution," op. cit., p. 8.
77. Fletcher, "The Ethics of Genetic Control," op. cit., p. 45.
78. Cited in Donald Fleming, "On Living in a Biological Revolution," *Atlantic,* vol. 223, no. 2 (February, 1969), p. 66.
79. Rick Gore, "The Awesome Worlds Within a Cell," *National Geographic,* vol. 150, no. 3 (September, 1976), p. 383.
80. James F. Danielli, "Artificial Synthesis of New Life Forms," *Bulletin of Atomic Scientists,* December, 1972, p. 22.
81. J. B. S. Haldane, "Biological Possibilities in the Next Ten Thousand Years," in Gordon Wolstenholme, ed., *Man and His Future* (Boston: Little, Brown, 1963), p. 354.

82. Roderic Gorney, *The Human Agenda* (New York: Simon & Schuster, 1972).
83. Cited in Halacy, op. cit., p. 146.
84. Charles Weigel and Stephen Tinkler, "Eugenics and Law's Obligation to Man," *South Texas Law Journal,* 14 (1973), p. 373.
85. Frankel, op. cit., p. 33.
86. Lappé, "Moral Obligations," op. cit., p. 416.
87. *Britannica Book of the Year, 1964* (Chicago: Encyclopedia Britannica, 1964), p. 499.
88. Leon R. Kass, "Making Babies—the New Biology and the 'Old' Morality," *The Public Interest,* Winter, 1972, pp. 53–54.

5
BIO-FUTURES

1. Harris Poll, 1969.
2. Joseph Fletcher, *The Ethics of Genetic Control* (Garden City, N.Y.: Anchor Books, 1974), p. 76.
3. Frederick Ausubel, Jon Beckwith, and Karen Janssen, "The Politics of Genetic Engineering: Who Decides Who's Defective?", *Psychology Today,* June, 1974, p. 34.
4. Fletcher, op. cit., p. 29.
5. Quoted in A. Rosenfeld, *The Second Genesis* (New York: Vintage Books, 1975), p. 182.
6. Fletcher, op. cit., p. 199.
7. Garland E. Allen, "Science and Society in the Eugenic Thought of H. J. Muller," *BioScience,* 20 (1970).
8. Julian Huxley, "Eugenics: an Evolutionary Perspective," *Perspectives in Biology and Medicine,* 6 (Winter, 1963), p. 23.
9. Gordon Taylor, *The Biological Time Bomb* (New York: Mentor, 1968), p. 180.

10. D. S. Halacy, Jr., *Genetic Revolution* (New York: Harper & Row, 1974), p. 151.
11. James Bonner, "Beyond Man's Genetic Lottery," in *The Next Billion Years: Mankind's Future in a Cosmic Perspective* (Moffett Field, Calif.: Ames Research Center, NASA, 1974), p. 64.
12. Caryl Rivers, "Genetic Engineering Portends a Grave New World," *Saturday Review,* April 8, 1972, p. 23.
13. A. Rosenfeld, op. cit., p. 287.
14. David Krech, "The Chemistry of Learning," *Saturday Review,* January 20, 1968, p. 48.
15. "Turning Prescriptions Into Profit," *Science for the People,* vol. IX, no. 1 (January-February, 1977); Carole Wade Offir, "Are We Pushers for Our Own Children?", *Psychology Today,* December, 1974; Sidney Walker, III, "We're Too Cavalier About Hyperactivists," *Psychology Today,* December, 1974.
16. "New Deodorant Works by Anesthetizing Noses," *Moneysworth,* March 14, 1977, p. 9.
17. Dr. A. M. Chakrabarty, "Which Way Genetic Engineering," *Industrial Research,* January, 1976, p. 48.
18. Chris Welles, "Warning: Cosmetics May be Hazardous to Your Health," *New Times Magazine,* June 25, 1976, p. 42.
19. Linda Wolfe, "Cosmetic Surgery: A New Look," *McCall's,* January, 1977, p. 138.
20. D. S. Halacy, Jr., op. cit., p. 184.
21. E. Kendall Pye, "Can We Live Forever?", *Saturday Evening Post,* March, 1977, p. 35.
22. Gene Bylinsky, "A Preview of the Choose Your Mood Society," *Fortune,* March, 1977, p. 220.
23. Daniel Zwerdling, "Pills, Profit & People's Problems," *The Progressive,* October, 1973, p. 46.
24. "Turning Prescriptions into Profit," *Science for the People,* op. cit.

25. Ibid.
26. Ibid.
27. Ibid.
28. Daniel Zwerdling, op. cit., p. 46; Fidell, "Put Her Down on Drugs" (available from Know, Inc., Box 86931, Pittsburgh, Pa. 15221).
29. D. Larned, "The Selling of Valium," *Ms.*, November, 1975.
30. Daniel Zwerdling, op. cit., p. 45.
31. D. S. Halacy, Jr., op. cit., p. 102.
32. Dreitzel, cited in Navarro, "A Critique of Ivan Illich," *International Journal of Health Services,* 5, p. 360.
33. House Subcommittee on Oversight and Investigations, 1976, as reported in the *Washington Star,* April, 1977.
34. Statement of Dr. John Finklea, Director, NIOSH, before the Subcommittee on Labor, Senate Committee on Human Resources, reported on *CBS Evening News,* May 9, 1977.
35. Louis Timinck, "Next Job Application Blank May Include Your Genotype, Too," *Houston Chronicle,* April 4, 1975, p. 2.
36. Dow Medical Team Report, delivered before the New York Academy of Sciences, 1975, *Annals of the N.Y. Academy of Sciences,* 269 (December 31, 1975).
37. Ibid.
38. Barnett and Mueller, *Global Reach* (New York: Simon & Schuster, 1975), p. 303.
39. Sam Zagovia, "Searching for Meaning in Work," *Washington Post,* February 6, 1972, p. B4.
40. U.S. Department of Health, Education and Welfare, *Work in America* (Cambridge: MIT Press, 1973).
41. Robert Layer, "Behavior Modification as a Managerial Technique," *Conference Board Record,* January, 1975, p. 23.

42. Joel Makower, "Office Design," *Washingtonian Magazine,* March, 1977.
43. Ibid.
44. Ibid.
45. James Mitchell and Rolfe Schroeder, "Future Shock for Personnel Administration," *Public Personnel Management,* July-August, 1974, p. 266.
46. D. S. Halacy, Jr., op. cit., pp. 145–46.
47. Joseph Fletcher, op. cit., p. 170; Charles L. Weigel, Stephen E. Tinkler, "Eugenics and Law's Obligation to Man," *South Texas Law Journal,* 14 (1973), p. 374.
48. Gairdner B. Moment, "Andromeda Strain?", *BioScience,* 24 (September, 1974), p. 487.
49. S. E. Luria, "Modern Biology: A Terrifying Power," *The Nation,* October 20, 1969, p. 408.
50. Carl A. Larson, "Ethnic Weapons," *Military Review,* November, 1970, p. 10.
51. *New York Times,* July 23, 1975, p. C17; July 24, 1975, p. 1; September 9, 1975, p. 1; September 10, 1975. *Washington Post,* July 31, 1975, p. 1; August 10, 1975, p. A4.
52. Nicholas M. Horrock, "Colby Tells Senate Panel About CIA Poison Work," *New York Times,* September 17, 1975, pp. 1, 27; John Cummings, "War Test in Cities," *Washington Post,* December 22, 1976, pp. 1, 14.
53. Rosenfeld, op. cit., p. 166.
54. Julian Huxley, "Eugenics: an Evolutionary Perspective," op. cit., p. 178.

6
SCIENTISTS AND CORPORATIONS

1. Internal memorandum of Cetus Corporation, October, 1975.
2. Ibid.

3. Personal interview with Jonathan King, April, 1977.
4. Jon Beckwith, "Gene Expression in Bacteria and Some Concerns About the Misuse of Science," *Bacteriological Review,* vol. 34, no. 3 (September, 1970), pp. 222–27.
5. Ethan Signer, "Gene Manipulation: Progress and Prospects," 30th Symposium of the Society for Developmental Biology, *Academic Press,* 1974, p. 231.
6. Source: Scientists Institute for Public Information.
7. National Science Foundation, *R & D In Industry, 1974* (NSF 76–322), p. vi.
8. Ibid., p. 2.
9. Charles Schwartz, "The Corporate Connection," *Bulletin of the Atomic Scientists,* October, 1975, pp. 15–19.
10. Ibid., p. 17.
11. Act to Incorporate the National Academy of Sciences approved March 3, 1868.
12. Phillip Boffey, *The Brain Bank of America* (New York: McGraw-Hill, 1975), p. 54.
13. Ibid., pp. 20, 25, 26, 32.
14. Stewart L. Udall, speech before AAAS, December, 1970.
15. Boffey, op. cit., p. 59.
16. Ibid., p. 68.
17. "Men & Power," *Nature,* 230 (April 30, 1971), p. 549.
18. Boffey, op. cit., p. 69.
19. Ibid., p. 79.
20. James Goddard, "The Drug Establishment," *Esquire,* March, 1969, p. 114.
21. Boffey, op. cit., p. 82.
22. Ibid., p. 83.
23. Ibid., p. 168.
24. Sponsors listed on Official Program Brochure for

NAS Forum on Recombinant DNA Research, Washington, D.C., March 7–9, 1977.
25. *Business Week,* August 9, 1976, p. 67.
26. Goddard, "The Medical Business," *Scientific American,* September, 1973.
27. FDA, *Consumer,* December, 1975–January 1976, p. 9.
28. Cited in *Ciba-Geigy Journal,* No. 3–41975.
29. Study authored by Senator Kefauver, 1962.
30. Milton Silverman and Philip Lee, *Pills, Profits, and Politics* (Berkeley: University of California Press, 1974).
31. "Abbott Pleads No Contest In Adulterated Drug Case," *New York Times,* June 5, 1974.
32. "University, Firm Sued Over DES Test in Mothers," *Washington Post,* April 26, 1977, pp. 1, 14.
33. "The Costly Aftermath of a Poison Cloud," *Business Week,* October 11, 1976, pp. 32–33.
34. Morton Mintz and Jerry Cohen, *Power, Inc.* (New York: Viking, 1976), p. 191.
35. David Burnham, "AMA Aide Let Upjohn Use Letter To Sell Drug," *New York Times,* July 8, 1975.
36. Testimony before Senate Small Business Monopoly Subcommittee, February 6, 1973.
37. Mintz and Cohen, op. cit., p. 186.
38. "AMA Invests Millions in Drug Firms," *Washington Post,* June 25, 1973; "Drug Firms Gave AMA $851,000," *Washington Post,* July 1, 1975.
39. HEW News Release, October 14, 1976, p. 76.
40. Jeremy Rifkin, Larry Gordon, and Dan Smith, "DNA," *Mother Jones,* February-March, 1977, p. 26.
41. Ibid., p. 39.
42. Robert Nisbet, "Knowledge Dethroned," *New York Times Magazine,* September 28, 1975.
43. *World Almanac* 1975, p. 587; Council on Eco-

nomic Priorities, *Guide to Corporations: A Social Perspective* (Chicago: Swallow Press, 1974), p. 23.

44. Ethan Signer, "Gene Manipulation: Progress and Prospects," op. cit., p. 229.

7
AT THE CROSSROADS OF HUMAN HISTORY

1. Paul Samuelson, cited in Jeremy Rifkin, *Own Your Own Job* (New York: Bantam, 1977), p. 9.
2. National Urban Policy Roundtable, Discussion Paper No. 7, "Cities in Trouble," July 1, 1976.
3. Figures compiled by the Industrial Union Division, AFL-CIO, 1975.
4. Figures on alcoholism provided by the National Council on Alcoholism. Mental-illness statistics from *U.S. Fact Book* (New York: Grossett & Dunlap, 1977), p. 87. Crime statistics, ibid., p. 153. Firearms statistics, ibid., p. 160.
5. Joshua Lederberg, *Bulletin of the Atomic Scientists,* October, 1966, p. 5.
6. Robert L. Sinsheimer, "Genetic Engineering: The Modification of Man," *Impact of Science on Society*, XX (1970).
7. Frances Warshaw, "Gene Implantation: Proceed With Caution," *Science for the People*, 1976.
8. T. Bogue and S. Wolfe, "Trimming the Fat Off Health Care Costs," *Public Citizen,* 1976.
9. Daniel S. Greenberg and Judith E. Randall, "Waging the Wrong War Against Cancer," *Washington Post,* May 1, 1977, pp. C1, C4.
10. Personal interview with Jonathan King, April, 1977.
11. Jon Beckwith, "Recombinant DNA: Does the Fault Lie Within Our Genes?", paper presented

to the National Academy of Sciences Forum on Recombinant DNA Research, March 9, 1977, p. 6; Ethan Signer, "Recombinant DNA is No Miracle Cure," testimony before the Subcommittee on Science, Research and Technology, Committee on Science and Technology, U.S. House of Representatives, March 30, 1977.

12. Personal interview with George Wald, April, 1977.
13. Robert Nisbet, "Knowledge Dethroned," *New York Times Magazine,* September 8, 1975, p. 34.
14. Amitai Etzioni and Clyde Nunn, "The Public Appreciation of Science in Contemporary America," *Daedalus,* Summer, 1974, pp. 191–205.
15. Gordon R. Taylor, *The Biological Time Bomb* (New York: Mentor, 1969), p. 219.
16. E. Teller, cited in *Science for the People,* 3, 1, p. 10.
17. John Platt, "What We Must Do," *Science,* November 28, 1969, p. 1117.
18. Alvin Weinberg, "In Defense of Science," *Science,* 167 (January, 1970), p. 144.
19. Bernard Davis, "Prospects for Genetic Intervention in Man," *Science* (December 18, 1970), pp. 1279–83.
20. James Watson, testimony before the Panel on Science and Technology, U.S. House Committee on Science and Astronautics, 1971.
21. Taylor, op. cit., p. 223.
22. Erwin Chargaff, remarks made before the National Academy of Sciences Forum on Recombinant DNA Research, Washington, D.C., March 7, 1977.
23. Ethan Signer, "Recombinant DNA Will Not Cure What Ails You," testimony before the Subcommittee on Health and the Environment, House Committee on Interstate and Foreign Commerce, U.S. Congress, March 15, 1977.

24. Personal interview with Jonathan King, April, 1977.
25. Jon Beckwith, "Recombinant DNA: Does the Fault Lie Within Our Genes?", paper presented to the National Academy of Sciences Forum on Recombinant DNA Research, March 9, 1977, p. 6.
26. Robert Sinsheimer, "The Dawn of Genetic Engineering," address to the Genetics Society of America (August, 1975).
27. Personal interview with Ruth Hubbard, April, 1977.

INDEX